MATHEMATICS
for the General Reader

E. C. TITCHMARSH

DOVER PUBLICATIONS, INC.
NEW YORK

To the three J's

This Dover edition, first published in 1981, is an unabridged republication of the work originally published in 1959 as a Doubleday Anchor Book by Doubleday & Co., Inc., Garden City, N.Y.

International Standard Book Number: 0-486-24172-6
Library of Congress Catalog Card Number: 81-66586

Manufactured in the United States of America
Dover Publications, Inc.
180 Varick Street
New York, N.Y. 10014

Contents

NOTE: The Dover edition, while retaining the original page numbers 13 through 197, contains fewer pages of preliminary matter. Nevertheless no part of the author's original material has been omitted.

Chapter I

COUNTING

Little children easily learn to count. Very early in their lives they notice the existence around them of recognizable objects. As soon as they can speak, they learn to say the names of some of these. Almost at once they notice that some objects may be classed together, as being obviously of the same kind. In particular they notice the existence of pairs of objects, and learn to use the word "two." When I speak of two hands and two feet, the child realizes that the set of my hands has something in common with the set of my feet. When I turn on a light and then another light, the child says "two lights." This is the beginning of counting.

Soon other numbers, three, four, five and so on are learnt. The use of the word "one" probably comes later, the existence of single objects being at first too obvious to call for a special name. "Nought," the negation of the existence of any objects of a particular class, is a comparatively abstract idea, which only occurs to us when we are used to counting. Some ancient races had no symbol for "nought," which they did not think of as the same sort of thing as "one" or "two."

Older children learn the routine of counting up to quite large numbers. Beyond thirty or forty this must soon cease to have any particular meaning for them, but the rhythm of counting (twenty-one, twenty-two, twenty-three) makes it rather like saying very easy poetry. Children sometimes even count backwards to amuse themselves.

It soon becomes obvious that the process of counting can go on a very long way. I once overheard my children discussing the question, "What is the largest number?" One of them thought that it must consist entirely of 9's. The second thought that it must be possible to get it by using all the words "hundred," "thousand," "million," and whatever else there might be, in the most favourable way (the idea of repetition not being thought of). The third objected that one could never count as far as that, supposing apparently that to make it fair one ought to be able to count through all the numbers up to the largest. They all agreed that the subject presented serious difficulties, and passed on to other topics.

They did not ask me what the largest number was. In this they were undoubtedly wise, because I should not have been able to tell them. I should have been faced, like any other mathematician, with a serious dilemma. Either there is a largest number, and when we get to it we must stop; or we go on for ever, and the set of numbers is endless, or, as we say, infinite.

It might be said that, as all the numbers which are ever actually used or thought of individually form a finite set, we might as well confine our attention to such a set, and avoid the necessity of trying to think about infinite classes of numbers. Perhaps it would be possible to do this, but it would really make the practice of mathematics more difficult. Not only should we be condemned for ever to the trivialities of finite arithmetic, but almost every statement in mathematics would be limited by a condition that the numbers involved must not be too large. Of course in our minds there is no barrier to endless counting. However far we have got, we can always count one more.

Practically all mathematicians agree that there is no upper limit beyond which counting must cease; that is, they agree to regard the numbers which begin with one, two, three, ——, the primal elements of mathematics, as an infinite class. Such an agreement, or declaration, which is itself incapable of proof, but which is a necessary starting point for further thinking, is called an axiom. The axiom about the set of numbers going on for ever is called the axiom of infinity.

What are numbers?

To children, and probably to most other people, numbers are just the things we count with. They are words such as "two" or "five," which call up in our minds a familiar set of objects, such as the set of my hands, or the set of fingers on one hand. The number spoken relates a named set to one of these familiar sets; that is, it asserts that we could pair off each object of the named set with one of the objects of the familiar set. "Two lights" might mean that there is a light on my right hand and a light on my left hand. But couples are so often met with that the set of hands may be forgotten, and "two" just relates a new couple to all those which we have met before.

Generally, if we can pair off the members of one set with the members of another set, so that none of either set is left over, then the two sets must have the same number, whatever that may mean. Number must have a meaning such that it is true that I have the same number of fingers on each hand, and the same number of buttons as buttonholes on my waistcoat (with coats the situation does not seem to be so simple).

The question what numbers *are* has been much debated by philosophers, and they do not seem to have reached any agreement about it. There is nothing particularly surprising or distressing about this. It has been said that mathematicians are happy only when they agree, and philosophers only when they disagree. Philosophic doubts about the nature of number have never prevented mathematicians from getting on with their calculations, or from agreeing when they have got the right answer. So perhaps the situation is satisfactory to all parties.

One of the most famous attempts to define number was made by Bertrand Russell. I will quote, for example, what he says in his book *An Introduction to Mathematical Philosophy*. "We naturally think that the class of couples is something different from the number 2. But there is no doubt about the class of couples: it is indubitable and not difficult to define, whereas the number 2, in any other sense, is a metaphysical entity about which we can never feel sure that it exists or that we have tracked it down. It is therefore more prudent to content ourselves with the class of couples, which we are sure of, than to hunt for a problematical number 2 which must always remain elusive. . . . Thus the number of a couple will be the class of all couples. In fact, the class of all couples will *be* the number 2, according to our definition. At the expense of a little oddity, this definition secures definiteness and indubitableness; and it is not difficult to prove that numbers so defined have all the properties that we expect numbers to have."

This plausible-sounding definition of the number 2 actually raises many difficulties. For example, are we really sure what we mean by "the class of all couples"? Are we to admit physical objects only, or "all objects of all thought" (in pairs) as members of the class? If, as we must suppose, it is the latter, it seems that I can always add to the class by thinking of a fresh couple, and thus that I can create couples which you know nothing about. It is true that you can always test any couple, the existence of which I announce, to see whether it is one; but that removes the ultimate "2" from the class of couples to some test for couples, which is just what Russell seems to wish to avoid.

Another objection is that arguments based on the supposed existence of classes such as the class of all couples lead to certain famous paradoxes which appear amusing, but which are rather destructive to theories of this kind. One of these runs as follows. Some classes are members of themselves; for example, the class of all classes is a class, and so is a member of itself. Others, such as the class of all men, are not members of themselves (since a class is not a man). Consider now the class of all classes which are not members of themselves. Is it a member of itself or not? If it is a member of itself, then by the definition of the class to which it belongs, it is not a member of itself. This is a complete contradiction, which shows that there is something unsound in the attempt to manipulate classes in this way.

Arguments of this kind, in which we seem logically to go round in circles, suggest awkward questions about the class of couples. The class of couples, together with the class of triplets, are two classes, and so should belong to the class of all couples. In fact, classes seem in this way to breed in an alarming manner.

Russell's definition of a number as a class of similar classes is very ingenious, but the difficulties which it involves have never been entirely cleared up. There are other schools of mathematical philosophers, known as Formalists and Intuitionists, who have put forward rival theories to explain what mathematics really is. No doubt this problem will be much studied in the future.

The conclusion of all this seems to be that we must do without a simple and direct answer to the question, "What is a number?" This will not prevent us from doing mathematics. I am all in favour of an intelligent theory of number. It should add to the pleasure of mathematics, just as an intelligent theory of rigid dynamics should add to the pleasure of bicycling. But it is possible to pedal along without it.

Most mathematicians feel that mathematics does not really rest on what philosophers define it to be, but that it has in it a harmony which somehow carries it along. This view seems to be supported by M. Black, in his book *The Nature of Mathematics*. He says: "The title of 'The Foundations of Mathematics' which the philosophical analysis of mathematics has often received is therefore a misleading one if, taken in conjunction with these contradictions, it suggests that the traditional certainty of mathematics is in question. It is a fallacy to which the philosopher is particularly liable to imagine that the mathematical edifice is in danger through weak foundations, or that philosophy must be invited like a newer Atlas to carry the burden of the disaster on its shoulders."

The view put forward by some philosophers, particularly the Intuitionists, that large parts of mathematics rest on insecure foundations and should therefore be abandoned, has never been accepted by the general run of mathematicians. It is no doubt a mistake to regard philosophers as enemies who would destroy our precious possessions. We may take comfort from the following sentence in the same book by Black. "Philosophic analysis of mathematical concepts therefore tends to become a synthetic constructive process, providing new notions which are more precise and clearer than the old notions they replace, and so chosen that all true statements involving the concepts inside the mathematical system considered shall remain true when the new are substituted."

Perhaps we could regard numbers as a sort of medium of exchange, like money. Most people are really interested in the goods and services which the world offers, and to them money is only a symbol for these. But it is not a meaningless symbol. A system of barter, in which we do without money and merely exchange goods, would be very inconvenient, and practically impossible in a complicated society. So a system in which we reduce all mathematics to statements such as "I have more fingers than you have noses" would be too cumbrous to contemplate seriously. Numbers are symbols, and very useful and interesting ones.

Chapter II

ARITHMETIC

As long as we are content simply with counting, numbers are practically just identification marks, like "Oakdene" or "Mon Abri," attached to objects in a row. But we can do things with sets of objects, merge them together, or break them up into parts. This is the origin of arithmetic.

Two sets of objects are said to be added when we think of the set of objects belonging to either one or the other of the given sets. This new set is called their sum. For example, the set of my limbs is the sum of the set of my arms and the set of my legs. The numbers of these sets of objects undergo a corresponding addition, which we denote by the sign $+$ (plus). I have two arms and two legs, and four limbs. This (and all other similar additions) is represented by the formula $2 + 2 = 4$. Here the sign $=$, read as "equals," means that the process of calculation indicated on the left-hand side gives the same result as the process of calculation indicated on the right-hand side, or that the result is the number written on the right-hand side.

It is a common experience that it does not make any difference in what order we add sets of objects. This is reflected in formulae such as $2 + 3 = 3 + 2$. That such a rule always holds is called the commutative law of addition.

Another important rule is called the associative law; this is indicated by the formula

$$(2 + 3) + 1 = 2 + (3 + 1).$$

The bracket notation means that whatever process is indicated inside the brackets must be done first, and that then the brackets may be removed. Thus the left-hand side means that we are to add 3 to 2, and then 1 to the result. The right-hand side means that we are to add 1 to 3, and then the result is to be added to 2. Both the processes lead to the number 6, and that the results are the same is just what the whole formula means.

Rules such as the commutative law are laws which we command our numbers to obey, so that they shall represent certain processes which are usually carried out with physical objects. We could command them to obey different laws, for example that $2 + 3$ should not be equal to $3 + 2$, but to something different. They would do this provided that the proposed laws were not inconsistent with each other, but then of course the whole system would mean something quite different from what it ordinarily does.

Next suppose, for example, that I have three sets each containing four members, and that I then count all the members straight through as one set. There are twelve in all. This is multiplication. It corresponds to the multiplication of numbers expressed by the formula $3 \times 4 = 12$. The number obtained by multiplying two other numbers is called their product. There are commutative and associative laws for multiplication, as examples of which we may give

$$2 \times 3 = 3 \times 2,$$

and $$(2 \times 3) \times 4 = 2 \times (3 \times 4).$$

There is also a "distributive law" relating addition and

multiplication, exemplified by

$$3 \times (2 + 1) = (3 \times 2) + (3 \times 1).$$

The symbol $-$ is the inverse of $+$; $3 - 2$ means a number which, when it has 2 added to it, gives 3; or subtraction can be defined in terms of the "logical subtraction" of sets. Similarly $\div$, the symbol of division, is the inverse of $\times$; $4 \div 2$ means a number which, when multiplied by 2, gives 4.

Negative numbers.

A negative number is just the same thing as an ordinary number, except that it carries round with it the sign $-$. Thus $-1, -2, -3$ denote negative numbers; -2 means, count 2 and then, if there are ordinary numbers about too, subtract instead of adding.

This seems a simple enough idea to us, but it took a long time to get itself clear in people's minds. This is probably because an ordinary number, say 2, at once calls up a vision in our minds of two objects; and -2 seems to mean that two objects are doing something even less than not existing, a difficult situation to visualize. It is however quite unnecessary to perform this feat. It is perfectly easy to count and to give a rule of operation as well, and this is all that a negative number does.

One can of course think of negative numbers as representing liabilities, if ordinary numbers represent assets.

To distinguish them from negative numbers, the ordinary numbers are called positive numbers. It is usual to set out the whole array of positive and negative numbers, together with 0, as a row extending endlessly in both directions

$$\cdots -3, -2, -1, 0, 1, 2, 3, \cdots$$

The use of negative numbers enables us to attach a

meaning to formulae such as $2 - 3$, which otherwise we should have to avoid. It can now just mean -1.

To multiply a negative number by a positive number, we first multiply without thought of sign, and then preserve the minus sign; thus $2 \times (-3) = -6$. To agree with the commutative law, we must also say that $(-3) \times 2 = -6$. But this implies that multiplication by a negative number is equivalent to multiplication, together with *reversal* of sign.

If this is to be true also when one negative number is multiplied by another, it must mean for example that -3 multiplied by -2 is 6; that is, in multiplication, two minuses make a plus.

Greater and less.

If I start pairing off my fingers against a shilling's worth of pennies, I find that there are some of the latter left over. In such a case we say that the number of fingers is less than the number of pennies. The symbol $<$ is used to mean "less than"; thus $10 < 12$. The opposite to "less than" is "greater than"; and this is denoted by $>$; thus $12 > 10$. We use $\leqslant$ to mean "less than or equal to." Similarly $\geqslant$ means "greater than or equal to."

These symbols are easy to remember, since the bigger number is always at the bigger end.

We bring negative numbers and 0 into this scheme by thinking of them in the order written down above. Thus we say that -3 is less than -2, and we write $-3 < -2$; similarly $-1 < 0$, $-1 < 1$, and so on.

The scale of ten.

Nature has provided us with ten objects, our fingers, with which to compare other sets of objects. No doubt this is the origin of the use of ten as a basic number in counting. This number was so used by most of the ancient races whose records have come down to us, the Egyptians, the Babylonians, the Greeks and the Romans.

The system of numbers which we all use today is as follows. The first nine numbers in order are denoted by the symbols 1, 2, 3, 4, 5, 6, 7, 8 and 9. Nought or zero is denoted by 0. These are known as the Arabic numerals. After 9 we do not have any new symbols, but the next number, ten, is denoted by 10, the symbol 1 being one place to the left. The next number is 11, then 12, and all numbers up to 99 can be formed in this way. After this the next number is 100, the symbol 1 being moved two places to the left to denote ten multiplied by ten. Similarly 1000 means ten multiplied by ten multiplied by ten, and so on. This gives us a simple way of writing down numbers indefinitely.

This system has the great advantage that it requires only ten symbols to express all numbers. When it has been learnt in the first few cases, it can easily be used to any extent. It is known as the scale of ten, or the decimal system.

The old notations for numbers seem very cumbrous compared with this. The ancient Greeks usually used letters for numbers. Thus for example Α, Β and Γ meant 1, 2 and 3; Ι, Κ and Λ meant 10, 20 and 30, and Ρ, Σ and Τ meant 100, 200 and 300. Thus 321 would be ΤΚΑ. This is quite a convenient system for fairly small numbers, but it has to be elaborated very much to express large numbers, and it gives no idea how to go on *indefinitely*. A special system was invented by Archimedes in order to show Gelon, King of Syracuse, that it was not beyond the power of language to express the number of grains of sand with which the universe, on a reasonable estimate, could be filled. A very interesting account of these ancient systems of numbers is given by T. L. Heath, *A Manual of Greek Mathematics* (Oxford, 1931).

The Roman numerals, with I as the first number, V for 5, X for 10, C for 100, D for 500, and M for 1000, are well known, and are still sometimes used.

In these old methods of counting, though ten was a specially important number, no use was made of position to denote multiplication by ten, in the sense in which we use it now. The only exception to this seems to have been a notation used by the ancient Babylonians. They had a system of signs in which V meant 1 and < meant 10; thus << VVV VVV meant 26, and << << meant 40. But << VVV<< VVV<< meant $26 \times 60 + 40$, the displacement of the symbol for 26 to the left meaning that it was to be multiplied by 60. This is equivalent to counting in the scale of 60. It does not seem to be known why they attached such special importance to the number 60.

A relic of this system has come down to us in our method of measuring angles, and of counting time. This has come via the Greeks, who copied the Babylonian system of astronomy. We divide a complete turn into 360 parts, called degrees; each degree is divided into 60 parts, called minutes; and each minute into 60 parts called seconds (i.e., second-sixtieths). A similar system is of course used for time.

Ten is a very convenient number for the scale, somehow not too large or too small. It has two factors, two and five, which makes it easy to see when any number is divisible by these numbers. But other numbers, such as twelve, would do equally well. We English have a sort of traditional affection for the number twelve (twelve inches make a foot, twelve pence make a shilling). In this country the scale of ten is used for counting but not for measuring, so that problems about money, weights and so on involve endless tiresome arithmetic. In other countries this has been avoided by the adoption of the metric system of measurement, and similar systems for weights and other things.

Any number greater than one could be used as the basic number of the scale. In a way, the simplest choice would be the number two. In the scale of two, the only symbols which we should have to use would be 0 and 1. These would have the same meanings as before. The number two is then denoted by 10, the displacement of 1 one place to the left meaning, in this scale, that it is multiplied by two. Three is 11 (i.e. two and one). Four is 100 (1 displaced twice to the left is multiplied by two twice). Similarly the remaining numbers up to twelve are 101, 110, 111, 1000, 1001, 1010, 1011, 1100. The disadvantage of this system is that it takes so many figures to represent moderately large numbers.

Factors.

When we multiply two numbers together, we obtain another number; for example, $2 \times 3 = 6$. It is then natural to ask about the opposite process. If we are given a number, can it be obtained by multiplying two other numbers? And if so, what are they? It is soon found that there are various possible answers to this question. The number 6 is the product of 2 and 3; the number 12 is the product of 2 and 6, and it is also the product of 3 and 4. On the other hand, the number 5 is not the product of any two smaller numbers. It is only the product of 1×5, which is not at all interesting.

If a number can be expressed as the product of two other smaller numbers, these are called factors of it; thus 2 and 3 are factors of 6, and 3 and 4 of 12. We can find out whether a number is a factor of another number by dividing and seeing whether there is a remainder left. Thus on dividing 7 into 45, we take 7's away from 45 as often as possible, that is 6 times, and then 3 is left. This amounts to expressing 45 as $(6 \times 7) + 3$. Clearly 7 is not a factor of 45; but it is a factor of 42, since nothing is left after the division. In fact we find that $42 = 6 \times 7$.

Sometimes one of the factors of a number has factors of its own; thus $12 = 3 \times 4$, and $4 = 2 \times 2$. In this case we can express the original number as a product of more than two factors; thus $12 = 3 \times 2 \times 2$. There may even be several factors; thus $60 = 2 \times 2 \times 3 \times 5$.

One thing which emerges from all this is the special position occupied by those numbers which have no factors, such as 2, 3, 5 and 7. These are called *prime numbers.* It is easy to write down a great many such numbers. The next few are 11, 13, 17, 19, 23 and 29.

Now in some of the above cases in which we have expressed a number in factors, these factors are prime numbers. In such a case we say that the number is expressed as a product of prime factors. For example $6 = 2 \times 3$ and $21 = 3 \times 7$ are such expressions. In other cases, such as $42 = 6 \times 7$, one or more of the factors is not prime (here 6 is not prime). But now we can write $6 = 2 \times 3$, and so $42 = 2 \times 3 \times 7$. We have thus expressed 42 as the product of three prime factors.

These examples suggest that any number whatever can be expressed as a product of prime factors. The reader should try a few examples, and will find that it always works. Another example is the above factorization of the number 60. But there is more in it than this. In all simple cases such as we have considered, there is only one way of expressing any given number as a product of prime factors, apart from trivial variations like changing the order of the factors. For example 60 is equal to 6×10, and also to 4×15; writing $6 = 2 \times 3$, $10 = 2 \times 5$, we obtain $60 = 2 \times 3 \times 2 \times 5$, and writing $4 = 2 \times 2$ and $15 = 3 \times 5$, we obtain $60 = 2 \times 2 \times 3 \times 5$. So the result is the same (apart from the order in which the factors are multiplied) in whichever way we do it.

These calculations suggest strongly the following

result: a number can be expressed as a product of prime factors in one way only. This is true, and not so very difficult to prove, but it is not quite obvious. We are so used to thinking of a number as the product of its prime factors that we are apt to regard the result as quite obvious. In trying to prove it, the main difficulty is to avoid assuming the truth of something equivalent to what is to be proved. For example, is it obvious that the product of two primes cannot be equal to the product of two different primes? However, we must leave the matter to books which deal with this subject.

The factorization of fairly small numbers is quite easy. If a very large number is written down at random, or expressed by means of some formula, it is often very hard to find out whether it has any factors or not. In a sense there is no difficulty about it, because one would merely have to divide by all smaller numbers to see whether any of them left no remainder. It is just that this would take so long to do, that (if the number were very large) one could not finish it in a lifetime. The invention of electronic computers has made it possible to factorize large numbers far more quickly. Even so, the numbers which we can factorize remain limited.

There are some simple rules for factors. If the figure representing units is even, 2 is a factor. If all the separate figures added together give a number divisible by 3, then the original number is divisible by 3. There are a few more rules of this kind, but they do not take us very far. The general problem of finding the factors of very large numbers is unsolved.

Prime numbers.

The prime numbers can be regarded as the raw material out of which all other numbers are made up. All numbers can be expressed as products of primes.

Let us then think how to find the primes, and how many of them there are.

A simple method of finding the prime numbers, used since ancient times, is called the sieve of Eratosthenes. Suppose for example that we want to find all primes less than 100. Write out all numbers from 2 to 100 in a row. Now 2 is a prime, but no other even number is, since they are all divisible by 2. So cross out all the even numbers (4, 6, 8, ···) after 2.

The first number not crossed out is 3, which is then a prime. Now cross out all subsequent multiples of 3 (6, 9, ···). Some numbers, such as 6, will have been crossed out twice, but this does not matter.

Proceeding in this way, we come to the primes 5 and 7, and cross out multiples of them. All other numbers up to 10, and a great many others, are crossed out. Now look at the numbers not crossed out. *They are all the primes up to* 100.

Why are they primes? Because if a number less than 100 has two factors, they obviously cannot *both* be greater than, or equal to, 10. So one at least must be less than 10. But we have crossed out all numbers with prime factors less than 10, and so all numbers with any factors less than 10. Hence the numbers left are primes.

This "sieve" method, which can of course be used up to any limit, is a good wholesale method of finding prime numbers. But it does not tell us much about how many of them there are.

It was proved by Euclid, or some mathematician of his period, that the sequence of primes is endless; or that, as we say, there are an infinity of primes. In other words, however far we go along the sequence of the numbers, there are always more prime numbers beyond. The proof of this is very ingenious but it is really quite simple, once you have thought of it (or Euclid has thought of it for you).

Let us take a fair-sized prime number, for example 97

(this is the greatest prime revealed by the sieve for primes less than 100). We want to prove the existence of at least one still larger prime. To do this, consider the number

$$(2 \times 3 \times 5 \times 7 \times \cdots \times 97) + 1,$$

which is the product of all the primes up to 97, plus 1. Now this number itself may be a prime. I do not know whether it is or not, but, if it is, the fact that it is proves at once what is required. It is an example of a prime greater than 97. Next suppose on the other hand that the number written above is not a prime. Then it has prime factors. But none of these prime factors can be the same as any of the primes 2, 3, $\cdots$ up to 97; for on dividing by any such number we obviously get the remainder 1 (this of course was why we added on the 1). Hence any of the prime factors must be greater than 97, and this again proves the existence of primes greater than 97.

Of course we have merely taken 97 as an example. The argument does not depend in any way on this particular prime number, and could be applied equally well to any other. We must therefore conclude that Euclid's theorem is true.

In modern times mathematicians have spent a great deal of time and energy in investigating the distribution of prime numbers. Such problems are usually very difficult, and a great part of their interest lies in the ingenious methods which are required to solve them. I will mention only one such problem. A glance down a table of prime numbers shows that they often go in pairs, separated only by a single even number. For example, 11, 13; 17, 19; 59, 61; are such prime-pairs. It is an obvious suggestion that such pairs, like the primes themselves, form an endless succession. But no method is known of showing whether this is true or not, and it remains a completely unsolved problem.

Squares, cubes, indices.

By the square of a number we mean the product of the number by itself; and we denote it by affixing a little 2 to the top right-hand corner of the number. Thus $1 \times 1 = 1^2 = 1$, $2 \times 2 = 2^2 = 4$, $3 \times 3 = 3^2 = 9$, are the squares of 1, 2 and 3. By the cube, we mean a number multiplied by itself, and then by itself again; for this we use a little 3; thus: $1 \times 1 \times 1 = 1^3 = 1$, $2 \times 2 \times 2 = 2^3 = 8$, $3 \times 3 \times 3 = 3^3 = 27$ are the cubes of 1, 2 and 3.

The square of a number is also called the second power of the number, and the cube is called the third power. Similarly the product of four equal factors is called the fourth power, and so on. The notation for fourth and fifth powers is quite similar to that for squares and cubes; for example we write

$$2 \times 2 \times 2 \times 2 = 2^4 = 16,$$

$$3 \times 3 \times 3 \times 3 \times 3 = 3^5 = 243$$

and so forth.

The little number in the top right-hand corner, which says how many factors are to be multiplied, is called an index.

The introduction of indices gives us an opportunity of illustrating the difficulty of determining the factors of a given number. Fermat, a famous French mathematician of the 17th century, conjectured that all the numbers $2^2 + 1$, $2^4 + 1$, $2^8 + 1$, $2^{16} + 1$, $2^{32} + 1$, and so on, are prime numbers. Here each index is twice the one before, and the numbers increase very rapidly as we proceed. The first three are equal to 5, 17, and 257, and these are all prime numbers; and the next one is also a prime number. Fermat's conjecture obviously rested on these facts, but it was not a very good idea, because the next two such numbers are not primes.

Euler, a great German mathematician of the 18th century, proved that

$$2^{32} + 1 = 641 \times 6700417,$$

and in the 19th century it was proved also that

$$2^{64} + 1 = 274177 \times 67280421310721.$$

The labour involved in doing arithmetic on this scale can be imagined. In recent times still more of Fermat's numbers have been factorized.

Very large numbers can be expressed by writing a number with a moderately large index. According to Archimedes, the number of grains of sand which the universe could contain would not exceed 10^{63}; in the decimal notation this is 1 followed by sixty-three 0's. The idea of calculating the number of particles in the universe has been revived recently. According to Eddington, this number is $2 \times 136 \times 2^{256}$. Written out in full, this would be a number of eighty figures.

Axiom, proof, theorem, hypothesis.

In this chapter we have been able to illustrate the general course which mathematical writings take. They begin with certain axioms or primary assumptions, which are supposed to be agreed upon between the writer and the reader. Such for example are the laws governing addition and multiplication, and the axiom of infinity. These are, so to speak, the rules of the game.

The object of mathematics is to prove theorems, that is, particularly striking and important results which follow from the axioms. For example, the statement that "a number can be factorized in one way only," and "the set of prime numbers is endless" are important theorems.

These theorems are derived from the axioms by means of proofs; a proof is a chain of reasoning, each link of which should be obviously valid according to the axioms, but the final result of which may be far from obvious. Many of the proofs in mathematics are very long and intricate. Others, though not long, are very ingeniously constructed. Some theorems are capable of being proved in several different ways.

The proof that the set of prime numbers is endless is a good example. It is not long, but it involves an ingenious idea which catches the attention of anyone capable of being interested in mathematics.

In many cases, mathematicians first guess theorems, and afterwards supply the proofs. Such a guess is called a hypothesis. Naturally it requires much experience of mathematics to be able to put forward reasonable hypotheses. The power to make hypotheses which are both interesting and reasonable is a sign of mathematical originality. It leads to many advances in mathematics. As examples of hypotheses, we may take the statements "the numbers $2^2 + 1$, $2^4 + 1$, $\cdots$ are all primes," and "the number of prime-pairs is infinite." The former, due to Fermat, turned out to be false. The latter has never been either proved or disproved.

Chapter III

ALGEBRA

Representation of numbers by letters.

As an example of the sort of thing which mathematicians do with numbers, suppose we set ourselves the following problem: to calculate the difference between the square of any number, and the square of the number next after it.

In the first few cases, this is quite easy to do. We have

$$2^2 - 1^2 = 4 - 1 = 3$$

$$3^2 - 2^2 = 9 - 4 = 5$$

$$4^2 - 3^2 = 16 - 9 = 7$$

$$5^2 - 4^2 = 25 - 16 = 9$$

and so on. To go on very far with this would be extremely tiresome. Also, the results written down on the right-hand side suggest that some sort of rule is operating, and that, if we could find out what it was, we could by-pass a lot of arithmetic.

Let us examine one case more closely, to find out if possible what is really characteristic about it. Take for example the third case written down above. It can also be written as

$$(3 + 1)^2 - 3^2 = 7,$$

where $(3 + 1)^2$ means that the numbers in the bracket are to be added together, and then the result is to be squared. But actually we could do the squaring without doing the addition first. This would go as follows

$$\begin{array}{r} 3+1 \\ \underline{3+1} \\ 3+1 \\ \underline{3^2+\ 3\qquad} \\ 3^2+(2\times 3)+1 \end{array}.$$

To get the first row of figures under the line, I multiply each of 1 and 3 by 1; to get the second line, I multiply each by 3, and set out the results one step to the left, for convenience of addition. Then add, and we get the answer as written.

Our example now takes the form

$$3^2 + (2 \times 3) + 1 - 3^2 = 7.$$

This shows where the result really comes from. It is made up as $(2 \times 3) + 1$, the 3^2 terms cancelling out.

Now exactly the same argument could be used in any other case. For example in the next case we have

$$\begin{aligned} 5^2 - 4^2 &= (4 + 1)^2 - 4^2 \\ &= 4^2 + (2 \times 4) + 1 - 4^2 = (2 \times 4) + 1, \end{aligned}$$

and that gives the result, 9; and so in all the other cases.

I hope the reader agrees that it would be intolerable to have to write out separately a lot of different sums of this kind, each of which is really quite similar to all the others. The question arises whether we cannot find some wholesale method, by which we can in some sense do them all at once.

The method actually consists of using a letter, for example n, to denote each of a class of numbers. For example, in the above case let n denote in turn each of the numbers 1, 2, 3, and so on. At any moment we can of course focus our attention on one particular number, e.g., 20, and say "let n be 20."

If n denotes any number, the next number is $n + 1$. The squares of these numbers are written n^2 and $(n + 1)^2$. Our problem then takes the form, what is the value of $(n + 1)^2 - n^2$, where n has any of the values 1, 2, 3, $\cdots$. To solve it, we proceed exactly as in the case $n = 3$. To calculate $(n + 1)^2$, we write

$$\begin{array}{r} n+1 \\ \underline{n+1} \\ n+1 \\ \underline{n^2+\ n\qquad} \\ n^2+(2\times n)+1 \end{array}.$$

Subtracting n^2, what is left is $(2 \times n) + 1$. The result can be written in the form

$$(n + 1)^2 - n^2 = (2 \times n) + 1.$$

In products involving letters, the sign of multiplication $\times$ is usually omitted. Thus we write $2n$ instead of $2 \times n$, and we should write the above formula as

$$(n + 1)^2 - n^2 = 2n + 1.$$

It is easily seen that this formula gives the above arithmetical results as particular cases, when we take n to be 1, 2, 3 or 4. But now any number of other cases can also be derived from it; for example $86^2 - 85^2 = 2 \times 85 + 1 = 170 + 1 = 171$.

The branch of mathematics in which classes of numbers are denoted by symbols, such as letters, in the above way, is called algebra. The expression just written down is an example of an algebraical formula. It is equivalent to a whole class—usually (as in this case) an *infinite* class of arithmetical formulae. Not only so, but it exhibits what is really characteristic about the arithmetical formulae. The algebra goes to the root of the matter, and ignores the casual oddities of particular cases.

In many formulae we use two or more letters at the same time. The classes of numbers from which the letters are to take their values may be the same or different. For example, we might say "let a and b each denote any positive integer." A formula involving such a and b is for example

$$(a + b)^2 = a^2 + 2ab + b^2.$$

This can be proved in exactly the same way as the above particular case, in which a was 3 and b was 1.

This is actually the simplest of a chain of formulae, which are together known as the binomial theorem. The next such formula is an expression for the cube of the sum of two numbers, and is

$$(a + b)^3 = a^3 + 3a^2b + 3ab^2 + b^3.$$

There are some kinds of algebra which are so complicated that they use up the whole alphabet, both capital and small letters, and the Greek alphabet too. A few years ago a mathematical paper was published in which a Chinese character was used as one of the algebraical symbols. Actually the possibilities are endless, since we can also attach suffixes to letters as additional labels, and write for example, $m_1, m_2, \cdots m_n, \cdots$ as algebraical symbols.

A few letters have come to be used in a different way, so that they always (unless there is a local rule to the contrary) mean the same number. The letter e and the Greek letter π are used in this way to denote specially important numbers.

Factors in algebra.

As an example of algebra, let us multiply together the two expressions $a + b$ and $a - b$. I call these expressions rather than numbers, because, though they represent numbers, they do not just do that. When I write "$a + b$" or "$a - b$" I mean "think of any two numbers you like, and then think of their sum or difference." The particular numbers thought of will not matter in what follows. This is really the essence of algebra, that the particular values of such expressions rather fall into the background; it is the way in which the numbers occur in the expressions that is of interest.

The multiplication goes in the same way as in the previous section; we can write it as

$$\begin{aligned}(a - b)(a + b) &= a(a + b) - b(a + b) \\ &= a^2 + ab - ba - b^2 \\ &= a^2 - b^2,\end{aligned}$$

the ab and $-ba$ cancelling out.

Hence the product of $a - b$ and $a + b$ is $a^2 - b^2$. Conversely, we can say that the factors of $a^2 - b^2$ are $a - b$ and $a + b$. This is really just another way of saying the same thing.

It is important to notice that the problem of finding the factors of $a^2 - b^2$ in algebra is quite different from that of finding the factors of particular values of this expression in arithmetic. Consider for example the factors of $5^2 - 4^2$, i.e. $25 - 16 = 9$. The algebra gives the formula

$$5^2 - 4^2 = (5 - 4)(5 + 4).$$

This is correct arithmetic too, but in arithmetic it has not got us anywhere, because in fact $5 - 4 = 1$ and $5 + 4 = 9$, so that the first factor is trivial, and the second can itself be factorized again. From the point of view of algebra, this is just an accident due to the particular values of a and b chosen. In algebra, $a - b$ and $a + b$ are the ultimate factors of $a^2 - b^2$. You cannot factorize $a + b$ any further, though you can, as it happens, factorize $5 + 4$.

The algebra books are full of examples of factors of

algebraic expressions. Here we shall recall just a few of them, which the reader should be able to verify quite easily. We have

$$a^3 + b^3 = (a + b)(a^2 - ab + b^2)$$

$$a^3 - b^3 = (a - b)(a^2 + ab + b^2)$$

and $$a^4 - b^4 = (a - b)(a + b)(a^2 + b^2).$$

None of these expressions can be reduced down any further into factors.

All these formulae are examples of algebraical *identities;* that is, they are equations in which the two sides are equal for all values of the numbers $a, b, \cdots$, and not because they have any particular values.

Inequalities.

As another illustration of algebra, let us consider what is meant by an inequality in algebra. In arithmetic, an inequality is just a relation between particular numbers, such as $2 < 3$. In algebra, an inequality involves letters such as $a, b, \cdots$ which represent whole classes of numbers, and an inequality is a relation of "greater than" or "less than" which is true for all numbers belonging to these classes.

As an example of an inequality in algebra, we shall show that *the square of the sum of any two numbers can never be less than four times their product.* Let us denote the two numbers by a and b. Then in symbols the theorem to be proved is $(a + b)^2 \geqslant 4ab$.

This is easily seen to be true in particular cases; for example $(2 + 3)^2 = 5^2 = 25$, which is greater than $4 \times 2 \times 3 = 24$; and $(3 + 4)^2 = 7^2 = 49$, which is greater than $4 \times 3 \times 4 = 48$. What is wanted however is a proof that the inequality is true independently of the particular values of the numbers concerned. Such a proof must depend merely on the way in which the a and b occur in the formula.

If one of the numbers, say a, is positive, and the other, b, is negative, the result is trivial. This is an expression often used by mathematicians, and it means that the result is so obvious that no formal proof is needed. The standard of triviality of course tends to alter as we become more experienced mathematicians. In this case, if a is positive and b is negative, the left-hand side of the proposed inequality, $(a + b)^2$, is positive or zero, and the right-hand side, $4ab$, is negative. The inequality then merely asserts that a certain positive or zero number is greater than a certain negative number, and this is trivial.

If a and b were both negative, the inequality would amount to the same thing as the corresponding formula involving $-a$ and $-b$ instead of a and b. Hence it is sufficient to consider the case where a and b are both positive.

To prove it in this case, we observe that

$$(a + b)^2 = a^2 + 2ab + b^2;$$

the difference between this and $4ab$ is $a^2 - 2ab + b^2$, and we have therefore to prove that this expression is necessarily positive or at least zero; and it is so, because it is equal to $(a - b)^2$, by the same rule for squaring an algebraic expression involving two numbers; and the square of any number, positive or negative, is positive. This gives the required proof.

We have been careful to write the inequality with $\geqslant$ instead of $>$, because it may happen that the two numbers $(a + b)^2$ and $4ab$ are actually equal. This will be so if the numbers a and b are equal, since each side of the proposed inequality is then $4a^2$; and a little consideration of the argument will show that this is the only case in which equality can occur.

As another example of an inequality, we shall prove that if a, A, b and B represent any four numbers, then

$$(aA + bB)^2 \leqslant (a^2 + b^2)\ (A^2 + B^2).$$

For example, if $a = 1$, $b = 2$, $A = 3$ and $B = 4$, then the left-hand side is $(3 + 8)^2 = 11^2 = 121$, and the right-hand side is

$$\begin{aligned}(1^2 + 2^2)(3^2 + 4^2) &= (1 + 4)(9 + 16) \\ &= 5 \times 25 = 125.\end{aligned}$$

There is no case in which this inequality is trivial, since both sides of it are obviously positive.

To prove it, multiply out both sides in full; it takes the form

$$a^2A^2 + 2aAbB + b^2B^2 \leqslant a^2A^2 + a^2B^2 + b^2A^2 + b^2B^2.$$

The difference between the right-hand side and the left-hand side is $a^2B^2 - 2aAbB + b^2A^2$, and by the formula for squares this is equal to $(aB - bA)^2$. As a square, this is necessarily positive, or at least zero. Hence the right-hand side is at least as great as the left-hand side, which was to be proved.

The subject of inequalities has occupied the attention of many mathematicians in recent times, and many remarkable results of the above kind have been proved.

Progressions.

As another example of algebraical formulae, we shall next sum the arithmetical and geometrical progressions. These will be required later in the book, and anyhow they are good examples of the way in which letters are used to represent whole classes of numbers.

A number is said to be the arithmetical mean of two other numbers, if it is equal to half their sum, i.e. if twice the first number is equal to the sum of the other two. Thus 2 is the arithmetical mean of 1 and 3, since $2 \times 2 = 1 + 3$; similarly 5 is the arithmetical mean of 3 and 7, and 10 of 8 and 12.

An arithmetical progression is a sum, such as $1 + 2 + 3 + 4$, or $3 + 5 + 7 + 9 + 11$, in which each term is the arithmetical mean of those on each side of it. It is of course easy to write down the values of these sums —they are in fact 10 and 35 respectively. But what we now want is a general rule by which the values of all such sums can be written down at once, however many terms they contain. As the simplest case of this problem, consider the sum

$$1 + 2 + 3 + 4 + \cdots + (n - 1) + n$$

containing n terms, where n is any positive number. The problem of summing this *for every* n is no longer a problem of arithmetic. It is one of algebra, since what we have to look for is a general formula expressing the sum of the progression in terms of n, valid whatever particular number n may represent.

This problem can be solved in the following way. Suppose first that n is an even number, i.e., n has 2 as a factor. Let $n = 2m$, say. Let us begin by adding the first term of the progression to the last; the result is $n + 1$. Next, add the second term of the progression to the last but one; the result is $2 + n - 1$, which is again $n + 1$; next the third term and the last but two make $n + 1$, and clearly this is a general rule. It is on the recognition of the existence of such general rules that proofs in algebra depend.

Now all the terms can be paired off in this way, since we have supposed that there is an even number of terms. The number of pairs is half the total number of terms, i.e., it is m; each pair is equal to $n + 1$; consequently the whole sum is equal to $m(n + 1)$.

This solves the problem if n is an even number. If n is an odd number, then $n - 1$ is an even number, and consequently the sum of the terms as far as $n - 1$ can be obtained from the above formula. Let $n - 1 = 2l$, say. Then the sum of the terms as far as $n - 1$ is ln. Hence the total sum is $ln + n = n(l + 1)$. The problem is therefore solved in all cases. For example, the sum of $1 + 2 + \cdots + 100$ is $50 \times 101 = 5050$, and the sum of $1 + 2 + \cdots + 149$ is $149 \times 75 = 11175$.

Any other arithmetical progression can be summed by using the formulae already obtained. For example $3 + 5 + \cdots + 25$ is equal to $3 + 3 + \cdots + 3$ (twelve terms) $+ 2(1 + 2 + \cdots + 11) = 36 + 2 \times 66 = 36 + 132 = 168$.

It is also easy to obtain general formulae for all such sums in a similar way.

There is another sort of progression called a geometrical progression. A number is said to be the geometrical mean of two other numbers, if its square is equal to the product of the other two numbers; for example, 4 is the geometrical mean of 2 and 8, and 6 is the geometrical mean of 4 and 9. The origin of the use of the words "arithmetical" and "geometrical" in the sense given here seems to be rather obscure, but the ideas involved are very simple.

A geometrical progression is a sum such as $1 + 2 + 4 + 8$ or $2 + 6 + 18 + 54$ in which each term is the geometrical mean of its two neighbours. The general form of a geometrical progression beginning with 1 and containing n terms is $1 + a + a^2 + \cdots + a^{n-2} + a^{n-1}$. The problem of summing the geometrical progression consists of finding a formula for this sum, depending of course both on n, the number of terms, and on a, the quotient of each term by the one before.

In this case it is no use adding pairs of terms, and a different device has to be thought of. Suppose that we multiply the whole progression by a (the same a, of course, as occurs in the progression). The result is $a + a^2 + \cdots + a^{n-1} + a^n$. Now subtract the original progression from this. The result is $a^n - 1$, since all the other terms cancel in pairs. Now what we have obtained in this way is $a - 1$ times the original sum. The value of the original sum is therefore $(a^n - 1) \div (a - 1)$. For example, $1 + 2 + 4 + 8$ is $(2^4 - 1) \div (2 - 1)$, i.e. $16 - 1 = 15$. Naturally the formula shows to more advantage when it is applied to longer sums, in which direct addition is not so easy.

The unknown x.

There is another way in which letters are used in mathematics to represent numbers. We sometimes write a letter instead of a number, not because we want to represent a whole class of numbers, but because we do not know what the number in question is. An unknown number is often represented by the letter x.

An equation is a formula which asserts that two numbers, arrived at by different processes of calculation, are in fact equal. An equation differs from an identity in the fact that it is not usually true for all values of the symbols which occur in it, but merely for some particular values of these symbols, or even for only one such value. It is then a question of finding out what these particular values are. This is known as solving the equation.

If the numbers on the two sides of an equation involve an unknown number x, to solve the equation is to find the value or values of x for which the equation is true. Such values are called the roots of the equation. Simple examples of equations are $2x + 3 = 11$, and $x + 1 = 2x + 4$. The technique of solving such equations is taught in algebra books. It can at once be verified that the solutions of these equations are $x = 4$ and $x = -3$ respectively.

Equations are often presented to us as practical problems. Suppose that I have 3d., and that two people each give me the same sum, and that then I have 11d. What sum did they each give me? Suppose that they each gave me x pence. Then the situation is precisely represented by the former of the above equations. Each, of course, gave me 4d.

Readers will probably remember being asked at school to solve problems of this kind: "A father is four times as old as his son. In twenty years time, he will be only twice as old. Find their ages." Here there are two unknowns, the ages of the father and of the son. Let us denote them by x and y. The first statement is then expressed by the algebraic equation $x = 4y$. After twenty years, their respective ages will have become $x + 20$ and $y + 20$. Consequently the second statement is expressed by the equation $x + 20 = 2(y + 20)$. These two equations are called "simultaneous equations," since they are both true for the same x and y. It can easily be verified that the solution is $x = 40$, $y = 10$. For the method of getting this I must again refer to algebra books.

Now consider the following problem: "A father is three times as old as his son. In ten years time the son will be twice as old as his father. How old are they now?"

This problem is obviously an idiotic one; but the algebra goes along quite happily. If the father's age now is x and the son's is y, the two statements are represented by the equations

$$x = 3y, \; y + 10 = 2(x + 10)$$

and the solution is $x = -6$, $y = -2$. The point is that the algebra is a machine which does just what it is asked to do, and no more. As we have forgotten to mention that there cannot be negative ages, the data are actually quite consistent, and the answer, though absurd if related to real life, is perfectly correct mathematically.

The theory of numbers.

This name is given to that part of algebra in which we ask questions about factors, the divisibility of certain numbers by other numbers, the possibility of expressing numbers by means of algebraic expressions of certain kinds, and other things of that sort. It is a very ancient subject, which was studied particularly by Diophantus, a mathematician of the 3rd century A.D. It has this peculiarity, that many of its problems are very easy to state, but very difficult to solve.

Some examples, involving prime numbers, have already been considered in Chapter II. As another example, consider the following problem: to find numbers x, y, and z such that $x^2 + y^2 = z^2$. This has an obvious connection with the famous theorem of Pythagoras, that the square on the hypotenuse of a right-angled triangle is equal to the sum of the squares on the other two sides. A well-known solution is obtained by taking $x = 3$, $y = 4$, and $z = 5$, since then $x^2 = 9$, $y^2 = 16$, $z^2 = 25$, and $9 + 16 = 25$. Another solution is $x = 5$, $y = 12$ and $z = 13$, since $25 + 144 = 169$.

What we ask in the theory of numbers is whether it is possible to manufacture such solutions indefinitely; or, what comes to the same thing, whether there is a general formula which always gives solutions. The result is as follows. Let a and b be any two positive numbers, of which a is the greater; and let $x = a^2 - b^2$, $y = 2ab$ and $z = a^2 + b^2$. Then the equation is satisfied, whatever the values of a and b. This is easily seen from the rule for squaring the sum of two numbers. We have $(a^2 - b^2)^2 = a^4 - 2a^2b^2 + b^4$ and $(2ab)^2 = 4a^2b^2$; adding, the result is $a^4 + 2a^2b^2 + b^4$, and this is $(a^2 + b^2)^2$. This proves the general rule. The 3, 4, 5 example is formed by taking $a = 2$ and $b = 1$, and the 5, 12, 13 example by taking $a = 3$ and $b = 2$. As another example, let $a = 5$ and $b = 2$. Then $x = 21$, $y = 20$ and $z = 29$; and it is easily verified that $21^2 + 20^2 = 29^2$.

This is merely an example of a large class of problems which were considered by Diophantus and other mathematicians.

This particular problem had a remarkable sequel. Fermat, the celebrated French mathematician already mentioned, was very much interested in these questions. He possessed a copy of Bachet's *Diophantus*, in the margin of which he noted some of his own discoveries or conjectures. One of these marginal notes asserted that it is impossible to solve the equation $x^n + y^n = z^n$ when n is any number greater than 2; that is, that there are no numbers x, y, z such that $x^3 + y^3 = z^3$ or $x^4 + y^4 = z^4$ and so on. This assertion has become known as Fermat's last theorem. The remarkable thing about it is that to this day no other mathematician has been able either to prove or to disprove it. It is not even known with certainty whether Fermat had a proof, or whether he was only guessing. As he made other conjectures, some of which have turned out to be false, we may think that it was a guess. If so, it was a remarkably good guess, since the theorem has been verified for a great many values of n, though not, as Fermat said, for all values.

Another theorem of Fermat, this time not very difficult to prove, is that, if p is a prime number, and a is another number not divisible by p, then $a^{p-1} - 1$ is divisible by p. For example, take $p = 7$ and $a = 2$. Then $2^6 - 1 = 64 - 1 = 63$, which is divisible by 7. On the other hand, if $p = 9$, which is not a prime number, and $a = 2$, then $2^8 - 1 = 256 - 1 = 255$, which is not divisible by 9.

Another celebrated theorem in the theory of numbers, due to Lagrange (1736–1813) asserts that every positive number is the sum of four squares. For example $5 = 2^2 + 1^2 + 0^2 + 0^2$, and $12 = 3^2 + 1^2 + 1^2 + 1^2$. The point of the theorem is that, however large the number concerned may be, it is never necessary to use *more* than four squares to express it. Compared with the proofs of many theorems in this subject, the proof of this cannot be called very difficult, but it is beyond the scope of this book. This theorem has suggested many others of the same kind, involving cubes, fourth powers and so on.

Chapter IV

FRACTIONS

In the previous chapter we gave some examples of equations containing an unknown x, the value of which had to be found from the equation. It happened that in each case there was an x which satisfied the equation; for example $2x + 3 = 11$ is satisfied by $x = 4$. But there are plenty of equations which cannot be satisfied in this way; examples are $2x = 1$, $2x = 7$, and $3x = 5$. Thus there is no number which when doubled, gives 1 or 7, or which, when multiplied by 3, gives 5.*

This is a serious situation, not only because we like to be able to solve equations, but because in real life we are familiar with "halves," "quarters" and so on, and we think that there ought to be something in mathematics which would enable us to think accurately about such things.

Since an equation like $2x = 1$ is not satisfied by any of the numbers already used, we must, if we are to do anything with it, invent a new system of numbers, with different properties, by which it can be satisfied. Fortunately we do not have to make a completely fresh start; we can make up our new numbers out of the old ones.

We now consider, not just single numbers, as before, but pairs of numbers, which might be written together

* The insolubility of the equation $3x = 2$ was remarked on by Gilbert, *The Gondoliers*, Act II; "One can't marry a vulgar fraction."

in a bracket, as for example (2, 3) or, in algebra (a, b). Each such pair is now thought of as a single element of our new system, and might be called a "complex number," or simply "a number" in a new sense. Such numbers are of no use until we define the rules by which they are to be used. We must say what addition and multiplication mean in the case of such numbers. Of course the "addition" and "multiplication" must have some connection with ordinary addition and multiplication, or we should not call them by these names. We could make up a considerable variety of rules, but most of them would lead to nothing of any interest. Those actually adopted are as follows. The sum of the numbers (a, b) and (c, d), which is written $(a, b) + (c, d)$, is defined to be $(ad + bc, bd)$. Thus for example $(2, 1) + (1, 3)$ is $(7, 3)$. The product of the numbers (a, b) and (c, d), which is written $(a, b) \times (c, d)$, is defined to be (ac, bd). Thus for example $(2, 3) \times (1, 5)$ is $(2, 15)$.

The multiplication rule is thus a very simple one. To obtain the product we multiply the two first components, and also the two second components of the given numbers. The addition rule is less obvious, and its purpose only appears later.

We say that the two numbers (a, b) and (c, d) are equal, and write $(a, b) = (c, d)$, if $ad = bc$. Thus for example $(3, 6) = (1, 2)$. It might be supposed that they would only be equal if $a = c$ and $b = d$. This is a special case of "equality," but it turns out to be inconvenient to restrict the definition to this case.

We must now examine the properties of this system of numbers. Consider first those numbers whose second components are 1, such as (2, 1) and (3, 1). For such numbers the rules of addition and multiplication reduce to

$$(a, 1) + (c, 1) = (a + c, 1)$$

and

$$(a, 1) \times (c, 1) = (ac, 1).$$

Further, the two numbers $(a, 1)$ and $(c, 1)$ are equal if, and only if, $a = c$.

We see that the numbers of this special class have exactly the same properties as the ordinary numbers which we were using before. In any relation involving ordinary numbers we may therefore just as well use the corresponding numbers of this special class. The number $(a, 1)$ does just as well as a, and for all practical purposes they can be treated as being the same thing.

Suppose now that we interpret the equations considered above, not as relations between numbers in the original sense, but as relations between numbers in the new sense which has just been explained. Consider for example the equation $2x = 1$. This is now interpreted to mean $(2, 1) \times x = (1, 1)$, where x is a "number" of the form (a, b). It is easily seen that a solution of this equation is obtained by taking x to be the number $(1, 2)$; for by the multiplication rule $(2, 1) \times (1, 2) = (2, 2)$, and, by the definition of equality $(2, 2)$ is equal to $(1, 1)$.

Any equation of the form $qx = p$, where p and q are any numbers (in the original sense) can be solved in the same way. We interpret it to mean $(q, 1) \times x = (p, 1)$; and the solution is $x = (p, q)$, since $(q, 1) \times (p, q) = (pq, q) = (p, 1)$.

We could develop the theory of these numbers in more detail, but perhaps it is unnecessary to do so. It will be seen that they have precisely the properties of the familiar fractions (vulgar fractions, as they are sometimes called). The number (a, b) is the fraction $\frac{a}{b}$, the first component a being called the numerator, and the second component b the denominator. We shall henceforth write them in this way. For convenience in printing, the fraction $\frac{a}{b}$ is also sometimes written in the form a/b. It is usually read as "a over b," or "a upon b." It is also sometimes read as "a divided by b" but of course division (in the original sense) is possible only if b is a factor of a.

One of my earliest memories of mathematics is of a feeling of being puzzled about fractions. I think this may have been due to being told that "a half is something which, when multiplied by 2, gives 1," without it being made clear that there was any such "something." This way of starting a mathematical subject, by laying down certain laws or axioms, which the objects of study are assumed to obey, is often used, particularly in geometry. It could no doubt be used to introduce new number-systems such as fractions, but is apt to give rise to a sense of mystery. Actually the line which we have taken here shows that there need be no mystery about fractions. They can just be built up out of the numbers which we are supposed to know about already.

Even when we have made up a new number-system by combining familiar numbers in a certain way, it is often felt that the new numbers are in some sense less "real" than the old ones. The words "real" and "imaginary" will be mentioned later in the book as being used in a different, but rather similar, sense. Once one of my children, when settled in for the night, said as a parting remark, "There isn't a real half of seven, is there?" Perhaps she felt that the problem of dividing seven into two equal parts could only be solved by a sort of fiction. This is of course true. But if we interpret such problems as we have done here, there is nothing imaginary about the solution.

The method which we have used, that of solving an apparently insoluble problem by re-interpreting it in terms of new numbers, will be used again later in other connections. It is indeed one of the principal sources of progress in mathematical ideas.

Fractions.

To distinguish them from the new numbers, which we call fractions, the old set are called whole numbers or integers. Thus we speak of positive integers and negative integers. There are also of course positive and negative fractions, a fraction a/b being negative if one of a and b is negative.

The rules which we have adopted are of course calculated to make fractions behave as we should expect them to. Consider for example the rule of equality, that $(a, b) = (c, d)$ if $ad = bc$, or, in the ordinary notation, that $a/b = c/d$ if $ad = bc$. This condition is satisfied if $a = kc$ and $b = kd$, where k is a positive integer. The rule thus asserts that $kc/kd = c/d$. This is just the ordinary rule of "cancelling," in which a factor k occurring in the numerator and in the denominator is removed from both.

Next consider the rule of addition, which, in the ordinary notation, is that

$$\frac{a}{b} + \frac{c}{d} = \frac{ad + bc}{bd}.$$

A simple explanation of this can now be given. By the rule of equality a/b is equal to ad/bd, and c/d is equal to bc/bd. This process is known as "bringing the two fractions to a common denominator." The expression "a common denominator" has passed into ordinary language, with a rather vague meaning of something common to various people or things. In mathematics it means simply the bd of the above argument. Having done this, we add by adding the new numerators, keeping the denominator the same. This gives the addition rule. The multiplication rule is

$$\frac{a}{b} \times \frac{c}{d} = \frac{ac}{bd}.$$

The rules for subtraction and division are of course arranged so as to make these processes the opposites of addition and multiplication. They are simply

$$\frac{a}{b} - \frac{c}{d} = \frac{ad - bc}{bd}$$

and

$$\frac{a}{b} \div \frac{c}{d} = \frac{ad}{bc}.$$

It is easily verified that the associative and commutative laws, and so on, extend to fractions; for example

$$\left(\frac{a}{b} + \frac{c}{d}\right) + \frac{e}{f} \text{ is equal to } \frac{a}{b} + \left(\frac{c}{d} + \frac{e}{f}\right).$$

The fraction $a/0$.

Fractions with denominator 0 are usually excluded from the scheme. There is nothing wrong or inconceivable about them, but their properties are rather inconvenient. The sum or product of any other fraction with such a fraction also has denominator 0. Also, by the rule of equality, any two fractions with denominator 0 are equal, and any fraction whatever is equal to the fraction $0/0$. To admit this would be too much. For example, it is true in general that two fractions, which are both equal to a third fraction, are equal to one another. This can easily be deduced from the rule of equality. It breaks down, however, if the third fraction is $0/0$.

The best thing to do about this is to exclude such fractions from consideration altogether; that is, when we refer to a fraction a/b, it is always implied that b is different from 0.

Numbers in order.

The whole numbers are usually thought of, not just *en masse*, but as occurring in a definite order; in fact they are visualized as they are usually written down, 1, 2, 3, 4, and so on. Here 2 is greater than, or beyond 1, 3 is greater than 2, and 2 lies between 1 and 3.

It is possible to extend these ideas to the whole set of fractions. Consider first fractions in which the numerator and the denominator are both positive. Then we say that a/b is less than c/d, and write $a/b < c/d$, if $ad < bc$. For example, $\frac{1}{2} < \frac{2}{3}$ since $3 < 4$. This scheme includes the integers, since we identify $a/1$ with the integer a. It is easily verified that, according to the definition, all fractions, in which the numerator is less than the denominator, are less than 1; all fractions in which the numerator is greater than the denominator but less than twice the denominator, are greater than 1 but less than 2, and so on.

Now consider three fractions $\frac{a}{b}$, $\frac{c}{d}$, and $\frac{e}{f}$ such that $\frac{a}{b} < \frac{c}{d}$ and $\frac{c}{d} < \frac{e}{f}$. Thus $ad < bc$ and $cf < de$. Hence

$$ad \times cf < bc \times de$$

and dividing by cd we obtain $af < be$. Hence $\frac{a}{b} < \frac{e}{f}$. In other words, if the first fraction is less than the second, and the second less than the third, then the first is less than the third. This is what we mean by saying that they come in a definite order.

Extension to fractions of algebraical formulae.

One of the advantages of algebra is that most of the formulae, once they have been proved for particular kinds of numbers (such as integers), can be extended automatically to new kinds of numbers, as soon as these have been invented. Consider for example the factor-formulae proved on page 38. These of course referred to integers, because those were the only numbers then at our disposal. But what is really characteristic about these formulae is their form, the way they fit together; and this comes from the rules of addition and multiplication of the integers, but not from the fact that they *are* integers. Consequently the proof and the form of the results are exactly the same, if we interpret the a, b, and so on to be either integers or fractions. For example $a^2 - b^2 = (a - b)\ (a + b)$ if $a = \frac{1}{2}$ and $b = \frac{1}{3}$, just as much as if $a = 5$ and $b = 4$.

As another example, consider the sums of the arithmetical and geometrical progressions. It is easily seen that, whether n is odd or even, the sum of the arithmetical progression $1 + 2 + \cdots + n$ can be written in the form $\frac{1}{2}n(n + 1)$. The only objection to doing this previously was that we had not yet attached any meaning to the fraction $\frac{1}{2}$. Of course the sum is necessarily an integer; either n or $n + 1$ is an even number, so that the factor 2 can be divided into one or other of them.

Next consider the geometrical progression. The formula which we obtained for the sum of this was

$$1 + a + a^2 + \cdots a^{n-1} = (a^n - 1) \div (a - 1).$$

If a is an integer, the right-hand side is of course an integer, the division being always possible. But we now see that the formula is true whether a is an integer or a fraction. No fresh proof of this is required; we have only to notice that the proof given in the case when a is an integer works just as well when a is a fraction. As an example of the formula in this new sense, we may take

$$1 + \frac{1}{2} + \frac{1}{4} + \cdots + \frac{1}{64} = \left(\frac{1}{128} - 1\right) \div \left(\frac{1}{2} - 1\right)$$

$$= \left(1 - \frac{1}{128}\right) \div \left(1 - \frac{1}{2}\right)$$

$$= \frac{127}{128} \div \frac{1}{2} = \frac{127 \times 2}{128} = \frac{127}{64}.$$

Later in the book we shall make some further extensions of what we mean by a number. Each time we do this, there will be a corresponding extension of the formulae of algebra; that is, the content of these formulae will be extended, while their form remains exactly the same.

That part of algebra which is known as the theory of numbers, on the other hand, does not extend to fractions and other such numbers. It involves ideas peculiar to integers, and it would really be better to describe it as the theory of integers or whole numbers.

General use of the "fractions" notation.

The notation $\frac{a}{b}$ or a/b has come into common use as a substitute for $a \div b$, whether a and b are integers or fractions; $a \div b$ means generally a number c such that $bc = a$. If a and b are integers, this is just the fraction $\frac{a}{b}$. If a and b are themselves fractions, say $\frac{m}{n}$ and $\frac{p}{q}$, $a \div b$ or a/b is equal to the fraction mq/np. Later we shall use the same notation with other systems of numbers.

"n *equations for* n *unknowns*."

The system of fractions enables us to solve many more equations besides those considered above. In the first place, as we have already seen, the equation $qx = p$, where p and q are any integers, has the solution $x = p/q$. But this is also true if p and q are any fractions, p/q being then interpreted as in the last section. For example, the solution of

$$\frac{2}{3}x = \frac{4}{5} \text{ is } x = \frac{\frac{4}{5}}{\frac{2}{3}} = \frac{3 \times 4}{2 \times 5} = \frac{12}{10} = \frac{6}{5}.$$

In all such cases there is one equation, and one unknown x whose value has to be found from it.

Next, we can solve systems in which there are two equations and two unknowns. An example of such a system which was soluble in integers was mentioned at the end of the last chapter. But now any such system with certain exceptions can be solved. For example, the system $x + 2y = 1$, $3x + y = 2$ has the solution $x = \frac{3}{5}$, $y = \frac{1}{5}$; and the system $\frac{1}{2}x + \frac{1}{3}y = 1$, $x - \frac{1}{4}y = \frac{1}{2}$ has the solution $x = \frac{10}{11}$, $y = \frac{18}{11}$. That these are the solutions is easily verified, but for the method of obtaining them I must refer to algebra books.

The general rule is that n equations of this form can be solved for n unknowns, but there are some exceptional cases, and the formulae are too complicated to be given here.

I will take an example from Minoan arithmetic.* The Minoans flourished in Crete in ancient times. Archaeologists have found many clay tablets with inscriptions in their language. For the most part this has not been deciphered, but some of the tablets obviously contain arithmetic, apparently accounts relating to various commodities. On three of the tablets, addition sums are worked out. These sums are as follows:

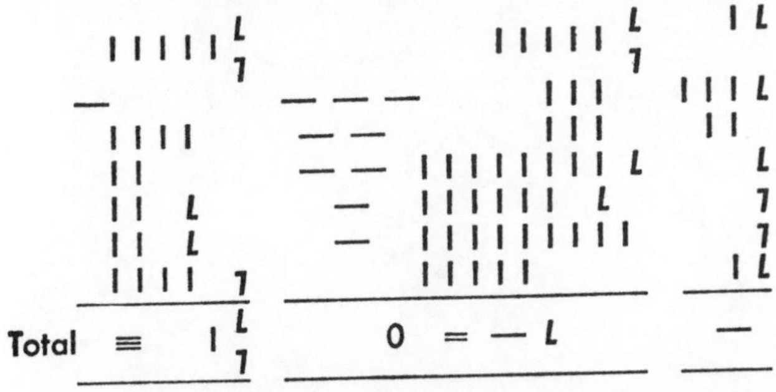

```
      IIIII L                  IIIII L      I L
            7                        7
  —            — — —             III      IIII L
      IIII       — —             III        II
      II         — —   IIIIIIIII L           L
      II  L        —   IIIIII  L             7
      II  L        —   IIIIIIIII             7
      IIII  7          IIIII                I L
  ___________  ________________________  ______
Total ≡   I L        0  =  — L              —
            7
  ___________  ________________________  ______
```

It is pretty clear that I is one, — is ten, and 0 is a

* I have to thank Sir John Myres for teaching me this subject.

hundred. The other symbols, **L**, **7** and $\frac{L}{7}$, must be fractions, and the problem is to find out what fractions they are. It has been conjectured that **L** is $\frac{3}{4}$, **7** is $\frac{1}{2}$ and $\frac{L}{7}$ is $\frac{1}{4}$. It is easily seen that this would make all three sums right. Actually the evidence is inconclusive. If **L** is x, **7** is y and $\frac{L}{7}$ is z, and we assume that the integers are correct, the three sums give us $2x + y + z = z + 2$, $2x + z = x + 1$, $3x + y + z = 3$. We have apparently three equations for three unknowns; but the third equation is merely a consequence of the other two. If we add the first two, side to side, we get

$$4x + y + 2z = x + z + 3.$$

On subtracting $x + z$ from each side, we get the third equation. Hence the third equation does not constitute independent evidence, and the solution is not completely determined. For example, another possible solution is $x = \frac{4}{5}$, $y = \frac{2}{5}$ and $z = \frac{1}{5}$. The conjectured values are probably the correct ones, but to settle the matter definitely still another worked-out sum will have to be found.

Chapter V

THE USE OF NUMBERS IN GEOMETRY

Geometry.

This is not supposed to be a book about geometry. This subject is to be dealt with in another book in the same series. But it is impossible to avoid saying something about it. Numerical ideas and geometrical ideas have been wrapped up together for thousands of years, and they cannot be separated entirely now. For the Greeks, geometry was the primary subject. A Greek writer pictured a product of two numbers as an area, and a product of three numbers as a volume. A formula in which four numbers were multiplied together was offered with a slight apology, as it did not correspond to any geometrical figure which could be visualized.

To-day, books which do not profess to deal with geometry are full of geometrical terms. Numbers are often called points, equations are called curves, and so on. We shall therefore say something about this subject. At the same time, we shall try to make our theories about numbers stand on their own feet. We shall use geometrical ideas as suggestions how to proceed, and as providing picturesque language, rather than as the real basis of our theories.

Geometry originally meant "land measurement." It arises generally from our attempts to give an account of the rigid non-interpenetrating objects with which we are familiar in the physical world. These objects force themselves on our senses, but it is not easy to think accurately about them. It is not simply that it is difficult to measure them exactly; it is impossible. Our measuring instruments, and indeed our senses, are coarse-grained. Objects which are sufficiently close together become indistinguishable, whatever aids to nature we employ in trying to separate them for observation. If we try to measure physical objects by dividing them up, we can do so up to a point; beyond this, they cease to be what they at first appeared. Divide a wooden foot-rule into twelve parts, and each is a wooden inch. Divide it into 10^{1000} parts and each is . . . what?

All this makes it very difficult to think in a reasonable way about the physical world. Actually we take refuge in thoughts about ideal systems which represent the physical world to our minds, but which, because they are constructed in our thoughts, are not subject to the awkwardness of physical objects. The name "geometry" has become transferred to the study of such ideal systems. Many such systems are very interesting in themselves, quite apart from the question whether they represent anything in physical space.

In geometry we consider things which we call "points," "lines" and so on. In our mental construction we may say that a point is anything which has certain properties which are required by the system, without specifying, or indeed being interested in the question what a point actually is. In this method of procedure, the question whether our assumptions are consistent, i.e., whether there could be *anything* with the assigned properties, is an important and sometimes a difficult one. But there is another way of thinking of geometry. We have already constructed one ideal system, that of numbers, and we can use them as the materials out of which to build our geometry. If we have already agreed about numbers, no further difficulties about existence arise; it is just a question whether numbers have properties which are interesting as a geometry. It turns out that this is so.

As the "points," the original elements of our geometry, we take the numbers, namely, the integers and fractions so far defined. This is reasonable because a "point" is an idealization of "a well-defined place," and a well-defined place is usually defined e.g., as "five yards along the path from where we are now." In fact measurement can hardly be done without counting.

The next step consists of describing properties of numbers by words which originally had a physical significance, and which therefore establish in our minds a correlation between our number-system and our ideas of the physical space which we intend it to represent.

"Large," "small," "near."

The idea of a large number is familiar, but the meaning of the word "large" depends on the context; 30 is a large score for a side at Rugby football, but not at cricket; 100 is a large individual score at cricket, but £100 a year is not a large income, and so on. In mathematics, we say that a number b is large compared with a number a if b can be divided into a large number of a's, with or without remainder. "Large" still has a rather vague sense; but it often refers to an indefinitely increasing scale of largeness—b is large compared with a, c is large compared with b, and so on.

"Small" is the opposite of "large"; a is small compared with b if b is large compared with a.

Transferring our attention to fractions, we say that the fraction $\frac{a}{b}$ is small if a is small compared with b. This still of course has the rather vague sense which is attached to all these words.

The difference between the fractions $\frac{a}{b}$ and $\frac{c}{d}$ is $\frac{a}{b} - \frac{c}{d} = \frac{ad - bc}{bd}$, and we say that $\frac{a}{b}$ is near to $\frac{c}{d}$ if $\frac{a}{b} - \frac{c}{d}$ is small. This use of the word "near" confers a geometrical meaning on our fractions. It is not inseparable from them, but I do not suppose that anyone actually thinks about fractions without visualizing them as marks on a ruler or in some geometrical way of this kind. Thus for purposes of thought and calculation we replace a ruler of unit length by the system of fractions $\frac{a}{b}$ with $a < b$. The particles of which the ruler is composed lie in a certain order along it, and so do the fractions lie in a certain order. Every particle has other particles very near to it on each side, and this is also true of the fractions, since $\frac{a}{b} - \frac{l}{n}$ and $\frac{a}{b} + \frac{l}{n}$ are very near to $\frac{a}{b}$ if n is very large. But of course the ideal system goes beyond the physical system in this matter of divisibility; by making n larger and larger we can make $\frac{a}{b} - \frac{l}{n}$ and $\frac{a}{b} + \frac{l}{n}$ as near as we like to $\frac{a}{b}$. We may thus say that the ideal system constructed of the fractions is *infinitely fine-grained.* As a representation of physical objects this is perhaps a defect, but a limitation on the fineness of the grain would introduce serious difficulties in calculation, just as a limitation on the largeness of numbers would.

Cartesian geometry.

The system of geometry in which a point is specified by means of a number, or is even identified with the number, was invented by the famous philosopher and mathematician Descartes. It is known as Cartesian geometry.

We saw above that the fractions correspond to the points along a ruler. If we think of all the fractions $\frac{a}{b}$, in which a may be less than or greater than b, or negative, these will correspond to a ruler extending to infinity in both directions. We call this system a one-dimensional Cartesian geometry.

According to this definition, the geometry is just the system of numbers; but it is intended to, and does, call up in our minds a picture of a sort of space, like the spaces of the physical world. There are problems of two different kinds about a system of geometry; whether just as a system it has interesting properties, and whether it corresponds to anything in physical space.

One-dimensional Cartesian geometry is not a very exciting subject; and a one-dimensional physical world would be rather lacking in interesting features. The inhabitants would be situated like beads on a wire or trucks on a railway-line. All one could do would be to move backwards and forwards. The behaviour of one's next-door neighbours would be even more important than it is in ordinary life.

Two-dimensional geometry.

A two-dimensional space is a space such as the floor of a room. Many people must remember constructing a sort of two-dimensional world on the nursery floor. The inhabitants are toys which are only allowed to move by sliding along the floor, and must not be lifted up. A hollow square of bricks lying on the floor makes a house, which the inhabitants can only enter or leave by opening a "door." If this system has three-dimensional aspects, we agree to ignore them. It is true that we can see down into the "house" from above, but from the point of view of the inhabitants we must be regarded as supernatural beings. If we throw something into the house from above, this must seem to have suddenly appeared inside a closed room, and must be regarded with horror by the inhabitants as a supernatural event.

Any place on the floor can be specified by measuring its distance from one wall, and also its distance from a second wall making an angle with the first. That is, two measurements are required. I am using the word "distance" here in a popular and rather vague sense. Later on this word will be given an official meaning in our theory. But it presumably conveys some idea to the reader, and we must be content for the moment to leave it at that.

The two measurements which specify a place on the floor give us two numbers, say x and y; that is, the point is specified by means of a number-pair (x, y). The numbers x and y are called the co-ordinates of the point. If we are building an ideal system, or geometry, in which points are numbers, the points of the system will *be* the number-pairs (x, y). These number-pairs must not be confused with those of Chapter IV, for which the bracket notation has now been abandoned. As we do not want to "add" or "multiply" points, there are no such rules of operation for these number-pairs.

The system consisting of the number-pairs (x, y) is called a two-dimensional Cartesian geometry.

This system is usually represented by a figure drawn on paper in the following way. A line, called the x-axis, is drawn across the paper, and another, called the y-axis, is drawn up and down the paper. These intersect at a point called the origin, and usually marked O. They are (mentally) extended indefinitely in both directions, and divide the paper (or plane) into four regions called quadrants. The number or point (x, y) is represented by a point on the paper distant x units to the right of the y-axis (or to the left, if x is negative), and y units above the x-axis (or below, if y is negative). Thus in the top right-hand quadrant both x and y are positive.

Straight lines.

I assume that the reader has learnt something about the "points" and "straight lines" in Euclidean geometry. A geometrical straight line is a mental construct which is supposed to represent the straight edges of physical objects. In Euclidean geometry, a straight line is really defined by the properties we assign to it.

In the Cartesian system we have identified "point" with "number," or "number-pair (x, y)," if we are in two dimensions. We must now consider what corresponds in this system to a straight line.

Similar figures.

One of the characteristic principles of Euclidean geometry is that of similar figures. Two triangles, for example, are said to be similar if their corresponding sides are proportional and their corresponding angles are equal.

If ABC and $A'B'C'$ are similar triangles, the former being the smaller, ABC is just a map of $A'B'C'$ on a smaller scale. In fact any small map is usually "similar" to the country of which it is a map. Of course in the case of a flat map of the earth, this is not strictly accurate, since the earth is round. The idea of similarity is really an extension to any size of the properties of scale drawings with which we are familiar.

There are systems of geometry in which similar figures on different scales do not exist. For example it can be shown that on the surface of a sphere there cannot be similar figures on different scales.

Equation of a straight line.

In the figure p. 66, APQ is a straight line intersecting the x-axis at A. Let P be (x, y), so that $OB = x$, and

$BP = y$. Similarly let Q be (x', y'), so that $OC = x'$ and $CQ = y'$.

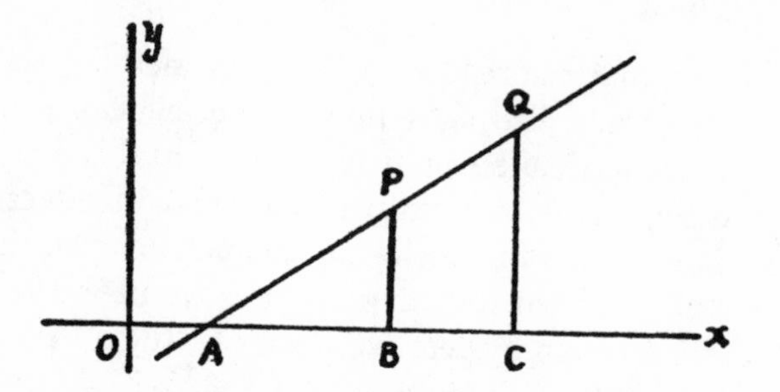

In this figure ABP and ACQ are similar triangles, and consequently corresponding sides will be proportional. Thus

$$\frac{BP}{AB} = \frac{CQ}{AC}.$$

If OA is of length a, then $AB = OB - OA = x - a$, and similarly $AC = x' - a$. The above relation is therefore equivalent to

$$\frac{y}{x - a} = \frac{y'}{x' - a}.$$

So if we took another point R on the line, of co-ordinates (x'', y''), we should find that $y''/(x'' - a)$ is also equal to the above expressions; in other words, $y/(x - a)$ has the same value for all points (x, y) on the line; it is a *constant*, i.e. it depends only on the position of the line, and not on the particular point $P, Q, R, \cdots$ chosen on it.

If it has the value c, then

$$\frac{y}{x - a} = c$$

or

$$y = c(x - a).$$

This relation is known as the equation of the straight line. It is, as we have seen, determined by the position of the line; and conversely, if the equation is given, the position of the line can be found.

If we think of an equation as something to be solved, well, we can solve it; i.e., if x is given we can determine y from the equation, or if y is given, we can solve it for x; the solution is

$$x = \frac{y}{c} + a.$$

It is easy to write down the equations of some special straight lines. The equation of the x-axis is $y = 0$. A straight line above the x-axis, everywhere at a distance 1 above it, is $y = 1$. These should be obvious from the definitions. Another example is the equation of the straight line which bisects the angle between the x-axis and the y-axis. Its equation is $y = x$; this expresses the fact that any point on the bisector is equidistant from the two axes.

From the point of view of algebra, the characteristic feature of the equation of a straight line is that it does not contain any squares or products of x and y, but merely constant multiples of x and y. We express this property by saying that the equation is *linear*. The general form of a linear equation is $lx + my + n = 0$, where l, m and n are constants. The constant l which multiplies the variable x is called the coefficient of x; and similarly m is the coefficient of y.

A straight line, in our system, must be defined officially to be what the above argument suggests. It is the set of all number-pairs (x, y) such that a linear equation $lx + my + n = 0$, with constant coefficients l, m, n, is satisfied.

We must point out also that two equations such as $x + y + 1 = 0$, and $2x + 2y + 2 = 0$, in which the coefficients and the constant term in one are proportional to those in the other, are thought of as corresponding to the same straight line. For the same value of x, each of them gives the same value of y, so that they correspond to the same figure in the diagram.

One often sees statements such as "a straight line has length but no breadth," and people sometimes wonder how anything can exist which is so lacking in solidity as to have no breadth at all. It is clear of course that such a statement cannot apply to the "straight lines" of the physical world. A line drawn on paper certainly has breadth, and even the boundary between two areas of different colours always has a certain vagueness if we examine it closely enough. In theoretical geometry, a straight line is just something we think of. In Cartesian geometry, what we think of is really a number. For example, one of the straight lines of the Cartesian plane is $x = 1$; and in the case of this line what it comes to is simply that any number is exactly equal to 1, or not equal to 1 at all. Two numbers are either the same or different. They may be nearly equal, but they do not merge imperceptibly into one another.

Distance.

We want to define something in our system of geometry which corresponds to the ordinary idea of distance in physical space. In one-dimensional geometry, the "distance" between the "points" represented by the numbers x and x' can be defined as being their difference, $x' - x$. This is reasonable, since if the points coincide the distance between them should vanish, while if they differ greatly it should be large. If we do not want to attach a directional meaning to the distance, we must take it to be $x' - x$ or $x - x'$, whichever is positive.

In two-dimensional geometry, that is, the geometry of number-pairs (x, y), we want to define the distance between the point P represented by (x, y), and the point Q represented by (x', y'). Consider first the case in which the two y's are the same, so that Q is (x', y). In this case, following the suggestion of one-dimensional geometry, it is natural to define the distance between P and Q to be $x - x'$ or $x' - x$, whichever is positive. If also $x = x'$ then the two points are identical, and the distance between them conveniently reduces to 0.

Similarly if the two points considered are (x, y) and (x, y'), with the same x, the distance between them is defined to be $y - y'$ or $y' - y$, whichever is positive.

To proceed to the general case, in which x is not equal to x' and y is not equal to y', we have to borrow some ideas from Euclidean geometry. We shall assume in particular that the reader knows the famous theorem of Pythagoras, about the sides of a right-angled triangle. We do this as a purely temporary measure, in order to suggest some interesting formulae. Later it is these formulae themselves which will provide our official definitions.

The theorem of Pythagoras is that the square on the hypotenuse of a right-angled triangle is equal to the sum of the squares on the other two sides. If PQR is such a triangle, with R as the right angle, this means that $PQ^2 = PR^2 + QR^2$. Now in Cartesian geometry, a straight line joining two points (x, y) and (x', y), with the same y, is usually thought of as being at right angles to one joining two points (x, y) and (x, y') with the same x. If P is (x, y), Q is (x', y'), and R is (x', y), then, according to this, PR will be at right angles to RQ.

If we are to reproduce in our Cartesian system Euclid's idea of distance and so also Pythagoras' theorem, we must therefore define the square of the distance between the points $P(x, y)$ and $Q(x', y')$ to be

$$(x - x')^2 + (y - y')^2.$$

This definition applies to any two points in the plane;

for example if $y' = y$, so that Q coincides with R, it reduces to $(x - x')^2$, as we should expect.

For the time being, we do not define "distance," but only "square of distance." The reader may think that this is a curious way to proceed, but the definition of distance presents difficulties which we are not yet in a position to surmount. If it makes the reader feel any better about this, we could call square-of-distance "separation" or something of the kind. We should then have defined the separation between two points, and its relation to the distance (if there is one) must be left over for the moment.

Parallel lines.

In Euclidean geometry, parallel straight lines are straight lines which never meet. We have to see what this corresponds to in Cartesian geometry. Suppose that the two lines are represented by the equations

$$lx + my + n = 0 \text{ and } l'x + m'y + n' = 0.$$

Then they are parallel if $l/l' = m/m'$, i.e., $lm' = l'm$. For example

$$x + 2y + 1 = 0 \text{ and } 3x + 6y + 2 = 0$$

are parallel.

To prove this, let us suppose that the two lines do meet; i.e. that there are numbers x and y such that both equations are true. Then we shall also have

$$l'(lx + my + n) - l(l'x + m'y + n') = 0.$$

On multiplying out, it is seen that the coefficient of x is zero, and so is the coefficient of y, in view of the relation $lm' = l'm$. All that is left is $l'n - ln' = 0$. But, if this is true, then l, m, and n are proportional to l', m', and n', so that the two lines are identical. If we exclude this case, the assumption that the two lines meet leads to a contradiction, and so cannot be true.

We can of course avoid an appeal to the principles of Euclidean geometry by taking the relation $lm' = l'm$ as the definition of parallelism. This in fact is the most convenient official definition in a number-geometry. But some such argument as the above is needed in order to show how the official definition arises.

Perpendicular lines.

Perpendicular straight lines are straight lines at right angles. But what is a right angle?

Practical men always knew what a right angle was. It was the angle of a tile such that four identical tiles would always fit together exactly on a flat floor, whichever way round you took them. In an ideal system it is, however, not quite so easy to define a right angle.

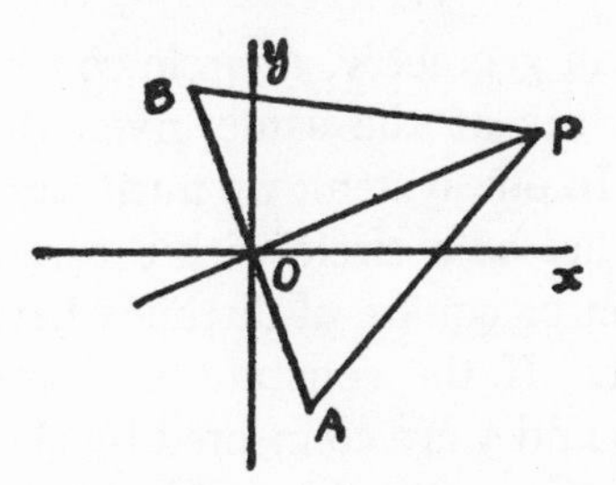

Again we shall borrow an idea from Euclid. In Euclid, if PO is perpendicular to AOB, AO being equal to OB, then $PA = PB$. Now let AOB be represented by the equation $lx + my = 0$, and let A be $(m, -l)$, and $B(-m, l)$. Let OP be represented by the equation $l'x + m'y = 0$, and let P be $(m', -l')$. Then the squared-distance PA^2 is $(m' - m)^2 + (l' - l)^2$, and the squared-distance PB^2 is $(m' + m)^2 + (l' + l)^2$. Written in full, these are $l^2 + l'^2 + m^2 + m'^2 - 2ll' - 2mm'$ and $l^2 + l'^2 + m^2 + m'^2 + 2ll' + 2mm'$ respectively, and the difference between them is $4ll' + 4mm'$. They are equal if this is zero, or, what comes to the same thing, if $ll' + mm' = 0$.

This can therefore be taken as our official definition of perpendicular lines. For example, $3x + 4y = 0$ is perpendicular to $4x - 3y = 0$. If we consider equations with a constant term, it is easily seen that this does not enter into the argument; for example, $3x + 4y + 1 = 0$ is perpendicular to $4x - 3y + 2 = 0$.

As a particular case, the lines $x = 0$ and $y = 0$, i.e., the "*axes*," are perpendicular, since they correspond to the values $l = 1, m = 0, l' = 0, m' = 1$. This does not necessarily mean that the axes must be represented by lines drawn on paper at right angles, in the popular sense. It is usual and convenient to do so, but, if we stick faithfully to our "official" definition, axes drawn in any other way ought to do equally well.

Circles.

In Euclidean geometry, a circle consists of all those points which are at the same given distance from a fixed point. In our system, we must say that the point P, or (x, y), lies on a circle with centre C, or (a, b), if the separation or square-of-distance between P and C is a constant. If the constant is denoted by k, this means that x and y are connected by the relation

$$(x - a)^2 + (y - b)^2 = k.$$

This is the equation of the circle. It corresponds in the Cartesian analysis to the circle, in the same way that the equations previously considered corresponded to straight lines. The constant k corresponds to the square of the radius in Euclid's system; the radius itself, like "distance," is left undefined for the moment.

There are many other interesting curves in plane geometry which correspond to simple equations connecting x and y. For example, the equation $y^2 = x$ represents a parabola, celebrated as the path of a projectile (though this parabola is lying on its side). The equation $x^2 + 4y^2 = 1$ represents an ellipse, celebrated as the orbit in which a planet moves round the sun. The equations $x^2 - y^2 = 1$ and $xy = 1$ represent hyperbolas. All these are the curves which have been studied since ancient times under the name of conic sections.

Three-dimensional geometry.

Life in three dimensions is perhaps too familiar to require description here. What we have to do is to represent by a mathematical system the space occupied by solid objects such as ourselves.

Consider an object in an ordinary room. Its position can be specified by stating the point on the floor which it is vertically above and its height above the floor. The point on the floor is fixed by two measurements from the walls, say x and y. Let its height above the floor be z. Then the position of the object is specified by the set of three numbers (x, y, z).

In solid or three-dimensional Cartesian geometry we therefore identify a point with such a set of three numbers. All the ideas of two-dimensional Cartesian geometry extend easily to three-dimensional geometry. Without going into details, we can say that a plane in solid geometry is represented by an equation of the form

$$lx + my + nz = c$$

where l, m, n and c are constants. The squared-distance between two points (x, y, z) and (x', y', z') is defined to be

$$(x - x')^2 + (y - y')^2 + (z - z')^2.$$

The equation of a sphere is of the form

$$(x - a)^2 + (y - b)^2 + (z - c)^2 = k.$$

Naturally much more can happen in three dimensions than in two, so that three-dimensional or solid geometry is more exciting, if more complicated than two-dimensional or flat geometry.

Is there a fourth dimension?

If it is true that poltergeists can throw solid objects into closed rooms, presumably they require a fourth dimension in which to do it. This is a situation with which (as mathematicians) we are perfectly prepared to cope. We have merely to add a fourth co-ordinate, and identify a point with a set of four numbers, say (x, y, z, w). This provides us with a four-dimensional Cartesian geometry. It is impossible to visualize it, but as a mathematical system it is not much more difficult to handle than three-dimensional geometry. In fact we can introduce any number of dimensions in the same way.

There is another way in which four-dimensional geometry has been used to represent physical space. We can represent time as a fourth dimension. The "points" of this geometry do not correspond to points in space, but to point-events. A point-event is specified by four numbers (x, y, z, t), where t is the time at which the event (supposed instantaneous) occurs. Thus (1, 2, 3, 4) would mean that something happened at the place (1, 2, 3) at 4 o'clock (if t represents the time in hours).

It might be supposed that the association of time with space in this way would serve no useful purpose, since time is usually thought of as something of quite a different kind from space; but in the formulae of relativity this is found not to be true, and it is "space-time" taken as a whole which the relativists always think about.

Chapter VI

IRRATIONAL NUMBERS

Intersections of straight lines, circles, etc.

We return now to two-dimensional geometry. In Euclidean plane geometry, we are specially interested in the points at which straight lines and circles intersect or meet. We have to consider what corresponds to "intersection" in Cartesian geometry.

Consider for example the two straight lines represented by the equations

$$x - y = 1, 2x - y = 3.$$

It is easily verified that each of these equations is satisfied if we take $x = 2$ and $y = 1$; that is, the point (2, 1) lies on each line, and is therefore their "point of intersection." Another example is

$$x - 2y = 1, 2x - y = 4,$$

which meet at the point $\left(\frac{7}{3}, \frac{2}{3}\right)$. In fact it is easily seen that any two straight lines meet in a point, unless the coefficients of x in the two equations are proportional to the coefficients of y. Such equations correspond to parallel lines, which, of course, never meet.

Now consider the question of the intersection of a straight line and a circle. Take for example the circle with centre at the origin and squared-radius 4. (Naturally the radius, when it is defined, will be 2.) The squared-distance of the point (x, y) from the "origin" $(0, 0)$ is, by definition, $x^2 + y^2$; and for every point on the circle this is equal to the square of the radius, i.e., it is 4. Thus the co-ordinates (x, y) of every point on the circle satisfy the relation $x^2 + y^2 = 4$; in other words, this is the equation of the circle.

Let us take as our straight line the bisector of the angle between the axes, which, as we have previously seen, is represented by the equation $y = x$. This line passes through the centre of the circle. If the mathematics of the situation is to correspond at all to the way in which we visualize it, or to the figures which we draw to represent it, the straight line and the circle certainly ought to intersect.

Now since $y = x$ on the straight line, at a point of intersection we must have $x^2 + x^2 = 4$, i.e., $2x^2 = 4$, so that $x^2 = 2$. It is therefore a question of finding a number x whose square is 2. Clearly such an x cannot be an integer, and we must suppose that it is a fraction, say $\frac{a}{b}$. We may suppose this fraction reduced to its lowest terms, i.e., that any factors common to a and b have been removed by cancelling; then a and b have no common factor. We thus want $\frac{a^2}{b^2} = 2$, i.e., $a^2 = 2b^2$.

Now $2b^2$ is an even number (a number of which 2 is a factor is called even, a number of which it is not a factor is called odd). Hence a^2 must be an even number, and so also a must be an even number, since the square of an odd number is clearly odd. So a has 2 as a factor, i.e., it is of the form $2c$, where c is another integer. On the other hand, b is an odd number, because we have supposed that a and b have no common factor, and so b cannot have the factor 2.

Our equation can now be written as $4c^2 = 2b^2$, and

hence, dividing each side by 2,

$$2c^2 = b^2.$$

But this shows that b^2 is an even number, and so also that b is an even number.

We have thus arrived at a complete contradiction; the argument shows in one way that b is an odd number, and in another way that it is an even number. Something has gone wrong. The only way back to sanity lies in abandoning the assumption that the equation $x^2 = 2$ has a solution. We must admit that it has not.

The result is that, in spite of appearances, the straight line and the circle never meet. Each finds its way through a gap in the other. The existence of such gaps shows that our number-system has grave defects as a way of representing the continuous objects or motions of physical space.

The whole matter can, of course be put into non-geometrical language. A number whose square is equal to a given number is called the square root of the latter. What we have just proved is that there is no exact square root of 2. Actually there is nothing exceptional about 2; 3, 5, 6, 7, 8, and in fact most other numbers have not got square roots.

Now it is very inconvenient to have to admit that there is no square root of 2. Not only does this lead to strange geometrical situations, but it would mean that in all sorts of algebraical problems we should have to distinguish between different cases, according to whether certain numbers had square roots or not. Practically speaking, if 2 has not got a square root it is necessary to invent one.

The square root of 2.

We have seen that it is impossible to solve the equation $\frac{a^2}{b^2} = 2$ with integers a and b. It is however possible to divide all fractions $\frac{a}{b}$, in which a and b are positive, into two classes, according to whether $\frac{a^2}{b^2}$ is less than 2 or greater than 2. Naturally every number of the former class will be less than every number of the latter class.

A division of all the numbers (i.e., integers and fractions) into two classes, as in the above case, every number of one class being less than every number of the other class, is called a section. The simplest way to attach a meaning to the square root of 2 is to define it as being *the section* defined above. It would do equally well if we defined it to be the lower class, i.e., the class of numbers a/b with $a^2 < 2b^2$, or the corresponding upper class, but there is no reason to prefer one to the other. Whether we call it a section or one class or the other only makes a slight difference to the language to be used, the situation in any case being substantially the same.

We denote the square root of a number x by $\sqrt{x}$. Thus for example $\sqrt{4} = 2$. This curious symbol was once an r, but it has become worn down by constant use.

Naturally the square root of any positive number can be defined in the same way as the square root of 2. For example, the square root of 3 is the section of all numbers a/b into two classes, such that $a^2 < 3b^2$ in the lower class, and $a^2 > 3b^2$ in the upper class. To be quite consistent, we ought even to define the square root of a number such as 4, which has a square root in the ordinary sense, in this way. The square root of 4 would be the section of all numbers a/b into two classes such that $a^2 < 4b^2$ in the lower class, and $a^2 > 4b^2$ in the upper class; or, what comes to the same thing, such that $a < 2b$ in the lower class, and $a > 2b$ in the upper class. This may seem a very cumbrous way of getting at something very simple, but it simplifies the logic of the situation.

It must be admitted that we have not been able to define $\sqrt{2}$ in so simple, or it may seem in so satisfactory a way as, for example, $\frac{1}{2}$; and it might be thought that, if there are gaps in the sequence of numbers, it would be more honest just to say so. Actually the definition which we have given does all that is required of the definition of a number. Consider for example the sum of the two numbers $\sqrt{2}$ and $\sqrt{3}$. With each of them there is associated a lower class of numbers in the above way, viz., the numbers $\frac{a}{b}$ such that $a^2 < 2b^2$, and the numbers $\frac{c}{d}$ such that $c^2 < 3d^2$. Now form the class of numbers $\frac{a}{b} + \frac{c}{d}$. These numbers are the lower class of a new "section," which is $\sqrt{2} + \sqrt{3}$. Similarly $\sqrt{2} \times \sqrt{3}$ is defined by the section of which the numbers $\frac{a}{b} \times \frac{c}{d}$ form the lower class.

We usually think of $\sqrt{2}$ as something having a definite numerical value; it is given in tables as

$$1.4142136 \cdots$$

Such "values" are apt to make us suppose that $\sqrt{2}$ is something more concrete than a "section," but this would be a delusion. All that the above "value" means is that the numbers

$$1,\ 1\frac{4}{10},\quad 1\frac{41}{100},\quad 1\frac{414}{1000},\ \cdots$$

are members of the lower class of the section defining $\sqrt{2}$; and that, if we go on far enough with this sequence (whatever it may be), we shall get as near as we like to members of the upper class.

Rational and irrational numbers.

It is convenient to have names by which to distinguish between the integers and fractions used so far, and numbers of the new kind which have just been introduced. We call the integers and fractions rational numbers, and all others irrational numbers. Thus $\frac{3}{5}$ is rational, $\sqrt{2}$ is irrational. The reader must get used to using words, with which he is familiar in other connections, as the technical terms of mathematics, of course with different meanings. It might be clearer if we put the technical terms in inverted commas and spoke of "rational" and "irrational" numbers; but it is too late to try to introduce this convention now. We can only ask the reader to insert inverted commas mentally for himself, wherever they are appropriate. I hope the reader will agree at any rate that "irrational" numbers are not irrational, in the dictionary sense of "not in accordance with reason."

The set of all the rational and irrational numbers form the complete set of numbers which are required for ordinary calculation. It might be thought for one awful moment that, if we operated again with irrational numbers in the same way as we did with rational numbers, we should come upon a new class of super-irrationals, and so on endlessly. But this is not so. Nothing fresh emerges from this process. The irrational numbers already defined are the only ones which exist.

Extension of Cartesian geometry to irrational numbers.

The extension of the idea of number to include irrational numbers leads to a corresponding extension of Cartesian geometry. Previously, a point in Cartesian two-dimensional geometry was a number-pair (x, y), where x and y were both rational. We now consider also as points number-pairs (x, y), where either x or y or both may be irrational.

This enables us to complete our geometry in a very satisfactory way. For example, in the problem of the intersection of the straight line $y = x$ with the circle $x^2 + y^2 = 4$, we can now say that they intersect at the point $(\sqrt{2}, \sqrt{2})$; they also intersect at the point $(-\sqrt{2}, -\sqrt{2})$. This corresponds exactly to what we should expect from a figure, in which the straight line and circle certainly look as if they intersected in two points.

It is now possible also to define the distance between any two points in Cartesian geometry. In two dimensions, the distance between the points (x, y) and (x', y') is

$$\sqrt{(x - x')^2 + (y - y')^2}.$$

It is the square root of what we previously called the "separation." For example, the distance between the point (1, 1) and the "origin" (0, 0) is $\sqrt{2}$. In the previous system this distance did not exist, because it was not a rational number. The introduction of irrational numbers enables us to make general statements about lengths and distances as we should like to, without concerning ourselves about the nature of their particular values.

In three dimensions, the distance between the points (x, y, z) and (x', y', z') is of course

$$\sqrt{(x - x')^2 + (y - y')^2 + (z - z')^2}.$$

We can even define distance in four or more dimensions in a similar way.

The discovery that irrational numbers are required to make geometry do what is expected of it was made by the ancient Greeks. It is one of their most important contributions to mathematics.

Extension of algebra to irrational numbers.

The whole apparatus of algebraical formulae extends at once to irrational numbers. For example, a formula such as $a^2 - b^2 = (a - b) \times (a + b)$, proved first when a and b are integers, and next when they are fractions, is now seen to be equally true when one or both of a and b are irrational. In fact the rules of operation with irrational numbers are just the same as those for operations with rational numbers, and such a formula is simply a consequence of these rules, and not of the particular nature of the numbers a and b.

As an example of the advantages of the use of irrational numbers in algebra, let us take the following theorem: *the arithmetical mean of any two positive numbers cannot be less than their geometrical mean.* The arithmetical mean of two numbers a and b is $\frac{1}{2}(a + b)$, and their geometrical mean is $\sqrt{ab}$. Expressed as a formula, the theorem is $\frac{1}{2}(a + b) \geqslant \sqrt{ab}$.

The very expression of such a theorem requires the use of irrational numbers; even if a and b are taken to be rational, $\sqrt{ab}$ will usually not be rational, so that the theorem has no meaning in the domain of rational numbers, unless a and b are connected in a special way so that $\sqrt{ab}$ is rational. Of course one can get rid of irrationals by squaring each side of the inequality, and replacing it by $\frac{1}{4}(a + b)^2 \geqslant ab$. This is in fact what we did in discussing inequalities in Chapter III. We proved there that $(a + b)^2 \geqslant 4ab$ for any two positive integers a and b. But now we can replace a and b either by fractions or by irrationals, and we can also divide by 4 and take the square root of each side. The theorem of the arithmetical and geometrical means as we have stated it then follows.

The theorem stated here, involving two numbers, is only a particular case of a theorem involving any number of numbers. Suppose that we are considering n numbers $a_1, a_2, \cdots a_n$. Then their arithmetical mean is $(a_1 + a_2 + \cdots + a_n)/n$, and their geometrical mean is the nth root of $a_1 a_2 \cdots a_n$ (i.e., the number whose nth power is $a_1 a_2 \cdots a_n$). The general theorem is still that the arithmetical mean is not less than the geometrical mean. This is a famous theorem of which several different proofs are known, but they are too complicated to be given here.

Chapter VII

INDICES AND LOGARITHMS

Indices.

We explained in a previous section that a little number ^{2} attached to the top right-hand corner of a number means that it is to be "squared," i.e., $x^2 = x \times x$. Similarly x^3 means "x cubed," i.e., $x \times x \times x$; and generally, if n is any positive integer, x^n (x to the nth power) means $x \times x \times \cdots \times x$, where n equal factors are multiplied together. Of course x^1 just means x.

The expression "to the nth power" has passed from mathematics into general use, as meaning "to a very great degree of intensity," or something of that kind. In mathematics, an nth power may be very large or very small. If x is greater than 1, then x^2 is greater still, x^3 greater again, and so on; the nth power of x will be very large, if n is very large. On the other hand, if x is less than 1, then x^2 is smaller still, and x^n will be very small if n is very large.

The only exception to this is the number 1. Obviously every power of 1 is equal to 1.

The main law governing indices is that, if two powers of a number x are multiplied together, the corresponding indices are to be added. For example

$$\begin{aligned} x^2 \times x^3 &= (x \times x) \times (x \times x \times x) \\ &= x \times x \times x \times x \times x = x^5, \end{aligned}$$

corresponding to $2 + 3 = 5$. This is obviously a general rule, and it can be written in the form

$$x^m \times x^n = x^{m+n}.$$

This suggests that it will be useful to give a meaning to negative indices, e.g., to expressions such as x^{-2}, x^{-3} and so on. If the rule just written down is to remain valid when n is negative, we must have for example

$$x^2 \times x^{-2} = x^{2-2} = x^0.$$

By ^{0} we mean "no factors are multiplied." This must mean 1, not 0; for if we took it to mean 0, we should always get 0 on multiplying it by anything else. Thus

$$x^2 \times x^{-2} = 1.$$

Hence x^{-2} must mean $\frac{1}{x^2}$; and generally, of course, x^{-n} means $\frac{1}{x^n}$.

We also use fractional indices. What is meant by expressions such as $2^{1/2}$, $2^{2/3}$ and so on?

The answer to these questions is again forced on us by the above law of indices. If it is to remain true when m and n are replaced by fractions, we must have for example

$$2^{1/2} \times 2^{1/2} = 2^1 = 2;$$

that is, $2^{1/2}$ must be the square root of 2, or $2^{1/2} = \sqrt{2}$. We can use the index $\frac{1}{2}$ as an alternative notation for the square root. But in other cases the index notation is much more useful than the "root" notation. Thus

$$2^{1/3} \times 2^{1/3} \times 2^{1/3} = 2^{1/3+1/3+1/3} = 2^1 = 2$$

i.e., $2^{1/3}$ means the cube root of 2; and

$$2^{2/3} \times 2^{2/3} \times 2^{2/3} = 2^{2/3+2/3+2/3} = 2^2 = 4,$$

so that $2^{2/3}$ means the cube root of the square of 2, or, what comes to the same thing, the square of the cube root of 2.

It is even possible to attach a meaning to expressions such as $3^{\sqrt{2}}$, where the index is an irrational number. To do so, we consider the class of numbers $3^{a/b}$, where a/b is the class of fractions such that $a^2 < 2b^2$. These numbers $3^{a/b}$, form the lower class of a new section, and it is by means of this section that the number $3^{\sqrt{2}}$ is defined.

Logarithms.

This celebrated aid to arithmetic was invented by John Napier, a Scottish mathematician (1550–1617), and put into a practical form by Henry Briggs (1556–1631), first Savilian Professor of Geometry at Oxford. It depends on the fact, which we have just noticed, that if you multiply two powers of a number x, you add the corresponding indices. Now addition is usually easier than multiplication. Let us see whether we can reduce all multiplication to addition.

It is first a question of choosing the number x. For purposes of practical calculation, this is always taken to be 10. Logarithms to the base of 10 are usually known as Briggian logarithms, though the advantage of using this base appears to have occurred independently to Briggs and Napier (see *A History of Mathematics* by F. Cajori). Actually Napier did not consciously use any definite base, but obtained his logarithms by a geometrical construction, which seems to be substantially equivalent to the definition of a logarithm by means of an integral explained later in this book. The idea of defining a logarithm as an index only became clear to mathematicians at a considerably later period.

A number such as 100 or 1000 can be specified by the power to which you have to raise 10 to obtain it. Thus 100 is 10^2, and 1000 is 10^3. This index is called the logarithm of the number to the base 10. It is denoted by the abbreviation log; thus log 100 = 2, and log 1000 = 3. Sometimes we want to remind ourselves what number is being used as the base, and then this written as a suffix: $\log_{10} 100 = 2$.

Now consider a number such as 2, which is not a power of 10. This cannot be expressed in the form 10^n, where n is an integer; but it can be expressed in this form by replacing n by a fraction or an irrational number. Of course it requires a considerable knowledge of mathematics to do this; but let us suppose that it has been done, and that $2 = 10^k$. Then k is the logarithm of 2 to the base 10, or, as we write it, $k = \log_{10} 2$. The actual value of this logarithm is about .301. Values of logarithms such as this are given in books of tables, and the whole advantage of the method depends on the previous calculation of these "log tables." Another example is log 3 = .477. These values are of course only "correct to three decimal places."

Now suppose that we want to multiply 2 by 3, and that we do so by means of logarithms. We have to add the logarithm of 2 to that of 3; the result is .778. Finally we have to find the number of which this is the logarithm. This also can be done by means of the same table, and it is seen that the result is very nearly equal to 6.

Naturally the method of logarithms does not show to advantage in such a simple case as this, in which the result is obvious anyhow. But if we had to multiply together several numbers, each with a large number of figures, it might save much time. It must be remembered of course that someone has spent a great deal of time in calculating the log tables; but once this has been done the results can be used over and over again.

The method of logarithms shows to even greater advantage in the calculation of square roots, cube roots, and expressions of that kind. There is a rather complicated arithmetical method of finding square roots, but (within the limits of accuracy given by the tables) the method of logarithms is much simpler. For more complicated powers it is practically the only method available. In the case of square roots it goes as follows. Suppose that we wish to find the square root of 3. From a book of seven-figure log-tables I find that log 3 = .4771213. Now the theory shows that log $\sqrt{3}$ is $\frac{1}{2}$ of log 3; hence log $\sqrt{3}$ must be approximately .2385606. Looking through the same table for the number of which this is the logarithm, I find that it is somewhere between 1.7320 and 1.7321. At any rate it is clear that the value of $\sqrt{3}$ to three decimal places is 1.732.

We must explain now what the particular advantage of the base 10 is. The above calculation could have been carried out equally well with any number as the base. This was because all the numbers involved were between 1 and 10. Suppose, however, that we want to use the logarithms of numbers outside this range, say between 10 and 100. What, for example, is log 20? In general (i.e., if the base is not 10) this would require a completely new calculation. But if the base is 10, we have $2 = 10^k$, where $k = \log_{10} 2$; and $20 = 10 \times 2 = 10 \times 10^k = 10^{k+1}$, by the rules of indices. Consequently $\log 20 = k + 1 = \log_{10} 2 + 1$, and in fact its value is 1.301 approximately. A similar argument applies to any number between 10 and 100. Hence if the logarithms of all numbers between 1 and 10 have been tabulated, the logarithms of numbers between 10 and 100 can be found just by adding 1, the decimal part remaining the same. Next the logarithms of numbers between 100 and 1000 can be found by adding 2; and in fact the logarithm of any number whatever can be found in a similar way.

As another example of the use of logarithms, consider the problem of the number of figures in Eddington's number 136×2^{257} referred to at the end of Chapter II.

From tables we find that $\log_{10}$ 136 is about 2.13, and that $\log_{10}$ 2 is .30103 to five decimal places. The logarithm of 2^{257} is 257 times this, and this is found to be 77.36 to two decimal places. Adding, we find that the logarithm of Eddington's number is approximately 79.5. Now the number whose logarithm to the base 10 is 79 is 10^{79}, i.e. 1 followed by seventy-nine 0's; the number whose logarithm is 80 is 1 followed by eighty 0's. Eddington's number is somewhere between the two, and therefore it is a number of eighty figures.

Chapter VIII

INFINITE SERIES

Sequences, limits.

There are many operations of mathematics in which we give our attention in succession to one number, then to another, then to a third number, and so on indefinitely. In ordinary counting, for example, we think of 1, 2, 3, and so on, where "and so on" means "it is clear how to proceed in this way indefinitely." Such a succession of numbers is called a *sequence.* Other examples are

$$1,\ 0,\ 1,\ 0,\ \cdots$$

$$1,\ \tfrac{1}{2},\ \tfrac{1}{3},\ \tfrac{1}{4},\ \cdots$$

$$1,\ \tfrac{1}{2},\ \tfrac{1}{4},\ \tfrac{1}{8},\ \cdots$$

where "$\cdots$" means the same as "and so on." More examples of such sequences will be given later.

In most cases it is in the ultimate behaviour of a sequence, the "and so on," that the main interest lies. The above examples show that there are various different possibilities. The sequence 1, 2, 3, $\cdots$ goes on increasing beyond all bounds; in technical language, we say that "it tends to infinity." The reader should not try to analyse this phrase too closely. It means just what we have said, and no more. It does not imply that there is a place "infinity" at which the sequence ultimately arrives.

The sequence 1, 0, 1, 0, $\cdots$ is said to oscillate. There is nothing particular to explain about this.

The terms of the sequence $1, \frac{1}{2}, \frac{1}{3}, \cdots$ become smaller and smaller as we go along; and in fact they become indefinitely small, passing on their way any small boundary which we like to set up beforehand. Such a sequence is said to "tend to the limit 0." The number 0 is not a term of the sequence, but it is that number to which the sequence points unmistakably. The sequence $1, \frac{1}{2}, \frac{1}{4}, \cdots$ obviously has the same property.

There is a technical notation for all this. For "tends to" we draw an arrow, $\rightarrow$; "tends to infinity" is written "$\rightarrow\infty$" (again ∞ should not be thought of as a number, but merely as a symbol occurring in this expression). We usually denote a typical integer by n (n for number). The property of the sequence $1, \frac{1}{2}, \frac{1}{3}, \cdots$ which has just been explained is then written as

$$n \rightarrow \infty, \frac{1}{n} \rightarrow 0$$

and read "as n tends to infinity, $\frac{1}{n}$ tends to 0."

We can of course easily make up sequences with limits other than 0; for example, the sequence $\frac{1}{2}, \frac{2}{3}, \frac{3}{4}, \frac{4}{5}, \cdots$ tends to the limit 1; in technical notation as

$$n \rightarrow \infty, \qquad \frac{n}{n+1} \rightarrow 1.$$

The relation of all this to the theory of irrationals is not made clear by these examples, because all the numbers and limits concerned happen to be rational. But there is a very important connection. The situation is briefly as follows. Suppose that we have a sequence of numbers $a_1, a_2, a_3, \cdots$ such that each is greater than the one before, but such that they do not increase beyond all bounds; suppose for example that we have a fixed number M which is greater than every number of the sequence. Then an important theorem says that the sequence does tend to a limit; i.e., that there is a number A such that $a_n \to A$ as $n \to \infty$. But this theorem depends entirely on the theory of irrational numbers. The number A may be irrational, and then the theorem would not be true if "number" merely meant "rational number."

Series.

In this subject too we think of a number, then think of another number, and so on, but now at each stage we add the new number to the total of those that came before. Suppose, for example, that the numbers concerned are $1, \frac{1}{2}, \frac{1}{4}, \cdots$. Then the operation is indicated by the formula

$$1 + \tfrac{1}{2} + \tfrac{1}{4} + \tfrac{1}{8} + \cdots$$

ending with the inevitable $+ \cdots$, or "and so on." This means that we first take the number 1; then take the number $\frac{1}{2}$, and form the sum $1 + \frac{1}{2}$; then take the number $\frac{1}{4}$ and form the sum $1 + \frac{1}{2} + \frac{1}{4}$, and so on. These sums are called the partial sums of the series. To consider the series amounts to the same thing as considering the sequence of its partial sums. Of course we cannot add *all* the terms of such a series. What we can do is to think about the sequence of the partial sums. If this sequence tends to a limit, the series is said to converge, or to be convergent. Otherwise, it is said to diverge, or to be divergent. In the case of convergence, the limit to which the partial sums of the series tend is called the "sum" of the series. It is in this case only that the series taken as a whole can be said to have a definite "value," namely its sum.

Let us examine the above example a little more closely. A geometrical illustration would probably help the reader in doing this. Think of the terms of a series as measurements made along a foot rule, graduated in inches. "1" means "one inch from the end." "$+\frac{1}{2}$" means "add another half-inch." This takes us half-way from the one-inch mark to the two-inch mark. "$+\frac{1}{4}$" means "add another quarter-inch." This again takes us half-way from where we were before to the two-inch mark. And in fact it is easily seen that the addition of every term of the series has precisely the same effect. We always get half-way towards 2, without ever getting quite there. However, the number 2 clearly has a special relation to this series, and in fact it is its sum in the sense which has been defined above. The result is expressed by the formula

$$1 + \tfrac{1}{2} + \tfrac{1}{4} + \tfrac{1}{8} + \cdots = 2.$$

The reader who has succeeded in following this argument to its conclusion has summed his first infinite series. Or is it the first? I seem to remember learning at quite an early age something about recurring decimals. Suppose that I try to express the fraction $\frac{1}{3}$ as a decimal. I must carry out the operation of dividing 3 into $1.000 \cdots$; 3 into 1 won't go; 3 into 10 goes 3 with remainder 1; so again I have to divide 3 into 10, and again it goes 3 with the remainder 1; and so on indefinitely. It is natural to write the result of the process in the form $.333 \cdots$, but in fact the division never really comes out to a final result. Of course $.333 \cdots$ is nothing more nor less than the infinite series

$$\frac{3}{10} + \frac{3}{100} + \frac{3}{1000} + \cdots$$

and it will be shown later that the sum of this series really is $\frac{1}{3}$.

Apparently it is possible to sum an infinite series without knowing it.

The notation which mathematicians use for series is as follows. We use $\sum$, the Greek capital sigma or "s," to mean "the sum of" whatever is written after it. Thus $\sum n$ means "the sum of all numbers such as n," belonging to some specified class. For example, by

$$\sum_{n=1}^{4} n$$

we should mean the sum of all the integers from 1 to 4, that is $1 + 2 + 3 + 4 = 10$. As another example, the reader may verify that

$$\sum_{n=1}^{3} n^2 = 14.$$

In the case of an infinite series, the symbol "∞" is written at the top of the $\sum$. The two infinite series which we have just summed would be written as

$$\sum_{n=1}^{\infty} \frac{1}{2^{n-1}} = 2$$

and

$$\sum_{n=1}^{\infty} \frac{3}{10^n} = \frac{1}{3}.$$

These two series are examples of what are called infinite geometrical progressions. We have already explained what a finite geometrical progression is; it is a sum of terms in which each term is the geometrical mean of its two neighbours. If the first term is 1, and the next term is x, such a progression is of the form $1 + x + \cdots + x^n$. Now we can go on adding such terms indefinitely, and if we do so we obtain the infinite series $1 + x + x^2 + x^3 + \cdots$, or $\sum_{n=0}^{\infty} x^n$.

It is easy to see that such a series can have no meaning if x is 1 or any larger number; for example, if x is 1, it is simply

$$1 + 1 + 1 + \cdots.$$

The partial sums are 1, 2, 3, $\cdots$ which just go on getting larger and larger; in other words, the series is divergent.

On the other hand, if x lies between 0 and 1 we can sum the series. To do this, we have to consider the behaviour of the "partial sums" $1 + x + \cdots + x^{n-1}$ when n is large. We have already obtained a convenient expression for this sum, namely $\dfrac{1 - x^n}{1 - x}$. Now, as we have explained in a previous section, the numbers x^n which occur in the numerator here become smaller and smaller, beyond all small bounds, when n is increased indefinitely; or, as we say, x^n tends to 0 as n tends to infinity. The limit of the right-hand side is therefore simply obtained by omitting this term; i.e., it is

$$\frac{1}{1 - x}.$$

The sum of the series to infinity is therefore $\dfrac{1}{1 - x}$, or, as we may write it,

$$1 + x + x^2 + \cdots = \frac{1}{1 - x}.$$

We have already had some examples of this formula. In the case $x = \frac{1}{2}$, the formula gives

$$1 + \tfrac{1}{2} + \tfrac{1}{4} + \cdots = \frac{1}{1 - \frac{1}{2}} = \frac{1}{\frac{1}{2}} = 2.$$

In the case $x = \dfrac{1}{10}$, it gives

$$1 + \frac{1}{10} + \frac{1}{100} + \cdots = \frac{1}{1 - \frac{1}{10}} = \frac{1}{\frac{9}{10}} = \frac{10}{9}.$$

This is equivalent to the formula used above for evaluating the recurring decimal $.333\cdots$ (to get it exactly, multiply throughout by $\frac{3}{10}$).

As another example of an infinite series which can be

summed, consider the series

$$\frac{1}{2}+\frac{1}{6}+\frac{1}{12}+\frac{1}{20}+\cdots$$

or as we may write it (to show the rule by which the terms are formed)

$$\frac{1}{1\times 2}+\frac{1}{2\times 3}+\frac{1}{3\times 4}+\frac{1}{4\times 5}+\cdots.$$

The nth term of this series is $\frac{1}{n(n+1)}$. Now it is easily seen from the rule of subtraction of fractions that this is equal to $\frac{1}{n}-\frac{1}{n+1}$. Applying this formula to each of the terms of the series, we see that the partial sum as far as the nth term is equal to

$$1-\frac{1}{2}+\frac{1}{2}-\frac{1}{3}+\frac{1}{3}-\frac{1}{4}\cdots+\frac{1}{n}-\frac{1}{n+1}.$$

Here all the terms cancel in pairs except the first and the last, so that the whole sum is equal to $1-\frac{1}{n+1}$. Now when n tends to infinity the term $\frac{1}{n+1}$ tends to 0. It follows that the series is convergent, and that its sum is 1.

There is a certain conclusion to which many people who have studied up to this point the subject of infinite series have jumped. Is it not true that, if the terms of the series tend to 0, the series is convergent? It seems as if this may be so, since the addition of small terms makes little difference to the sum. Actually the suggested theorem is false. The best-known example of this is the series of the reciprocals of all the integers,

$$1+\frac{1}{2}+\frac{1}{3}+\frac{1}{4}+\frac{1}{5}+\cdots.$$

To prove that this series is divergent we do not do anything particular with the first two terms, which are together equal to $\frac{3}{2}$. Now take the next two terms, $\frac{1}{3}$ and $\frac{1}{4}$. The latter is the smaller, so that we have two terms, neither of which is less than $\frac{1}{4}$. Their sum is therefore not less than $\frac{1}{2}$.

Next take the next four terms, $\frac{1}{5} + \frac{1}{6} + \frac{1}{7} + \frac{1}{8}$. Of this the last is the smallest. Consequently we have four terms, none of which is less than $\frac{1}{8}$. Their sum is therefore not less than

$$4 \times \frac{1}{8} = \frac{1}{2}.$$

At the next stage we take eight terms, each not less than $\frac{1}{16}$; and so on and so on. As many times as we like we can find a block of terms to add, the sum of which is not less than $\frac{1}{2}$. But if we go on adding $\frac{1}{2}$ and $\frac{1}{2}$ and $\frac{1}{2}$ indefinitely the total sum will increase beyond all bounds. We have thus proved that the series which we have been discussing is divergent.

This proof has some useful lessons for the beginner in mathematics. It shows, what the reader may by this time be ready to grant anyhow, that the truths of mathematics are often not obvious ones. We are constantly coming upon questions which are quite easy to put, but not at all easy to answer. It often requires a very ingenious argument to decide what the truth is. In this lies much of the fascination of mathematics. It is like an endless game against a skilled opponent. If we can think of the right move, we win. Once we have made the right move, we gain some definite piece of knowledge which is never afterwards in doubt.

How to think of the right move is another question. It is largely a matter of experience. Mathematical technique consists of the accumulated bright ideas of thousands of years. Let us examine our bright idea about the series $1 + \frac{1}{2} + \frac{1}{3} + \cdots$ to see what it really amounted to. It was like a gambit in chess, in which a piece is sacrificed to gain what turns out to be a winning advantage of position. We want to prove that the partial sums of the series are ultimately very large. We do it by replacing them by something smaller. I said "$\frac{1}{3} + \frac{1}{4}$ is greater than $\frac{1}{4} + \frac{1}{4} = \frac{1}{2}$." Of course we could simply add, and get $\frac{7}{12}$, which is larger still. The point is that, by simply adding, we get more and more complicated sums as we go along the series, until we are finally lost in hopeless arithmetic. But replacing each block of terms by the same number of terms, each equal to the last of the block, is something which we can go on doing endlessly without getting lost. It is a winning move.

History of infinite series.

Infinite series were used by many mathematicians before they had any clear idea of what we call convergence or divergence. It may seem surprising that it was possible to do this; but what the old-time mathematicians were really doing was to use infinite series as if they were finite sums, and since this is often justifiable they usually got the right answer. In other cases they obtained results which seem strange to us. One of Euler's papers contains the formula

$$\cdots + \frac{1}{x^2} + \frac{1}{x} + 1 + x + x^2 + \cdots = 0.$$

This is a series infinite in both directions. The "proof" simply consists of combining the formulae

$$x + x^2 + \cdots = \frac{x}{1 - x}$$

and

$$1 + \frac{1}{x} + \frac{1}{x^2} + \cdots = \frac{x}{x - 1}.$$

The objection to this is that, while each of these formulae is correct for some value of x, there is no value of x for which they are both correct; the former is true when x is less than 1, and the latter when x is greater than 1.

Some of the formulae used by writers of that time are not open to such obvious objections. The formula $1 - 1 + 1 - 1 + \cdots = \frac{1}{2}$ was used by Leibnitz and John Bernoulli, and Euler gave $1 - 3 + 5 - 7 + \cdots = 0$. These series are not convergent, according to our definitions. We might call them oscillatory. In the former, for example, the partial sums are alternately 1 and 0. These numbers have no definite limit, but the number $\frac{1}{2}$ does happen to be their average value. In modern times methods have been given of attaching a definite meaning to, or "summing" such series, according to which the above formulae are actually correct.

Decimals.

In the above sections we have referred several times to decimals. We must now put down our ideas on this subject in a more systematic way.

In an expression such as 23.781, the figures to the left of the dot, or decimal point, represent an integer, which has been sufficiently discussed. The first figure to the right of the dot, here 7, means $\frac{7}{10}$; the next to the right again, here 8, means $\frac{8}{100}$; the next, 1, means $\frac{1}{1000}$; and so on, if there are any more figures. Thus the above expression means $23 + \frac{7}{10} + \frac{8}{100} + \frac{1}{1000}$. Clearly this is a rational number.

We call such an expression a finite decimal. We can also think of infinite decimals, in which the sequence of numbers after the decimal point never terminates. An infinite decimal may be formed by endlessly repeating one figure, or a block of figures. Examples are .333 $\cdots$

or .141414 $\cdots$, or .23111 $\cdots$ in which repetition starts after a certain point. These are called recurring decimals. Other infinite decimals (such as those for π or e) are formed in a less obvious way.

Now an infinite decimal is just a particular kind of infinite series. We have already pointed this out in some simple cases, and the same thing holds in every case. Also it can be proved without difficulty that every infinite decimal is a convergent series, and consequently that it has a definite "sum." Thus every decimal, finite or infinite, corresponds to a certain number, either rational or irrational. There is also a simple rule to show which sort of number the sum is. If the decimal is finite or recurring, the sum is rational; otherwise it is irrational.

It can also be proved without much difficulty that the converse of this theorem is true; every number can be expressed as a decimal, finite or infinite; and for each number there is only one expression as a decimal, provided that we agree not to use recurring 9. The necessity of this arises as follows. Consider for example the decimal .0999 $\cdots$, where the 9 recurs. This is equivalent to the series

$$\frac{9}{100} + \frac{9}{1000} + \frac{9}{10000} + \cdots$$

or to

$$\frac{9}{100} \times \left(1 + \frac{1}{10} + \frac{1}{100} + \cdots\right).$$

The sum of the series in brackets has already been found; it is $\frac{10}{9}$. Hence the value of the decimal is $\frac{9}{100} \times \frac{10}{9} = \frac{1}{10}$. But $\frac{1}{10}$ can also be expressed as .1, and this is obviously a simpler and more natural way of doing it. The use of a recurring 9 is therefore superfluous and can be ruled out.

The fact that any irrational number can be thought of as a decimal, even if it does end with the inevitable "$+ \cdots$" gives the irrational numbers a sort of concreteness. We can operate with these decimals just as we do in ordinary finite arithmetic, and with use the irrationals soon acquire a familiarity equal to that of the rationals.

How many irrational numbers are there?

This must seem at first a senseless question, because you can make up as many irrational numbers as you like; as we say, there are an infinity of them. The point is that it is possible to distinguish between some kinds of infinite aggregates and others. Some are, so to speak, more infinite than others. This may seem a surprising and even nonsensical statement, but examples show that it can be given a definite meaning.

The standard infinite set is the set of positive integers $1, 2, 3, \cdots$. This is the material out of which our whole system has been built, and it is natural to compare any other infinite set with it. Given any other infinite set (say for example, the set of rational numbers, or the set of irrational numbers) we ask whether it is possible to number off the members of this set, so that each member corresponds to just one of the positive integers. Suppose for example that the set is the set of even numbers, $2, 4, 6, \cdots$. In this the typical number is $2n$, and if we make this correspond to the integer n, the "numbering off" is achieved at once.

A set for which this can be done is called an enumerable set.

Consider next the set of all rational numbers between 0 and 1. These also can be "enumerated," in the following way. Take first the number $\frac{1}{2}$, because its denominator, 2, is the smallest possible. Next take the numbers $\frac{1}{3}$ and $\frac{2}{3}$, their denominators being the next smallest, 3, and their numerators running through the numbers less than 3 in order. As fourth and fifth in the list take the numbers $\frac{1}{4}$ and $\frac{3}{4}$, for similar reasons, $\frac{2}{4}$ being omitted since it is equal to $\frac{1}{2}$. As sixth, seventh and eighth in the list we take $\frac{2}{5}$, $\frac{3}{5}$, and $\frac{4}{5}$, and so on, increasing the denominator by 1 each time, and then taking the numerators in order of size, omitting fractions already expressed in a simpler way. Clearly every fraction finds a definite place in this list, though it is not quite easy to write down the place taken by a given fraction m/n. It will be noticed, however, that we do not say "arrange all the rational numbers between 0 and 1 in order of magnitude, and count them off in that order." This would not work, because, if you take any particular rational number, it is impossible to fix on any other one as being the next greater. There would always actually be others in between.

We have thus shown that the set of the rational numbers between 0 and 1 is enumerable. Now consider the same problem for *all* the numbers between 0 and 1, both rational and irrational. In the case of the irrational numbers there are no denominators to start work on, so that the method used above is not available; and of course an order-of-magnitude arrangement will not help, any more than it does with the rational numbers. We are at a loss to know what to do.

In fact, the problem is insoluble. This is a remarkable result, first proved by the German mathematician Cantor. It can be proved as follows. Suppose that the desired result had been achieved, and that all numbers between 0 and 1 had been enumerated in a definite way. Since each such number can be expressed as a decimal, this amounts to an enumeration of all the decimals. For the sake of writing down something definite, suppose that the first three decimals in the list were

$$.3427\cdots$$

$$.2981\cdots$$

$$.0446\cdots$$

to four decimal places. Now the point is that, if we could make such a list of all the decimals, we could derive from this list another decimal, different from every one in the list. The first decimal in the list begins .3 $\cdots$. Very well, let the proposed new one begin .4 $\cdots$. The second one in the list has a 9 in the second place; very well, let the new one go on .40 $\cdots$. The third one has a 4 in the third place; very well, let the new one go on .405 $\cdots$. The rule of operation consists of adding 1 to the nth figure in the nth decimal, except that, to avoid a possible recurring 9, we replace 8 by 7, not by 9.

The new decimal thus constructed differs in the nth place from the nth decimal in the list, and so it is different from every one of them. The assumption that it is possible to arrange *all* the numbers between 0 and 1 in such a list is therefore proved to be false.

The theorem shows that the "infinity" of irrational numbers is of a different kind from the "infinity" of rational numbers. There are in a sense many more of them. If we can imagine a situation in which we could pick out a number at random from all the numbers between 0 and 1, it is infinitely improbable that this number would be rational. This of course is not the same thing as being asked to think of a number "at random." We should presumably think of one of the most familiar ones, and it would turn out to be rational.

Chapter IX

π AND e

The circumference of a circle.

Suppose that we want to measure the distance round a hoop, barrel or round object of any kind. One way would be to tie a string round it, so that the ends just meet, and then to pull the string out straight and measure that. Another way would be to place the hoop on the ground with a mark on it against a mark on the ground, and then roll it along until the mark on the hoop comes down again. The distance between the two points on the ground corresponding to the mark on the hoop would be the length round the hoop.

Now consider the problem of the length of a circle in Cartesian geometry. To roll a purely ideal circle along an entirely conceptional straight line is not so easy. In fact it is not obvious that there is any definite number associated with a circle which can reasonably be called its length. A different method of approach to this problem is required. What we can do is to construct inside the circle polygons which follow the line of the circle round very closely. The length of each side of a polygon is naturally taken to be the distance between its end-points, distance having been defined in Chapter V. The length of the perimeter of the polygon is then the sum of the lengths of its sides. We may then expect that the length of the perimeter of the polygon will be an approximation to the length of the circumference of the circle.

Let us consider a circle of radius 1. In Cartesian geometry such a circle is represented by the equation $x^2 + y^2 = 1$. First of all, inscribe in it a square represented by $ABCD$ in the figure. The point A is (1, 0)

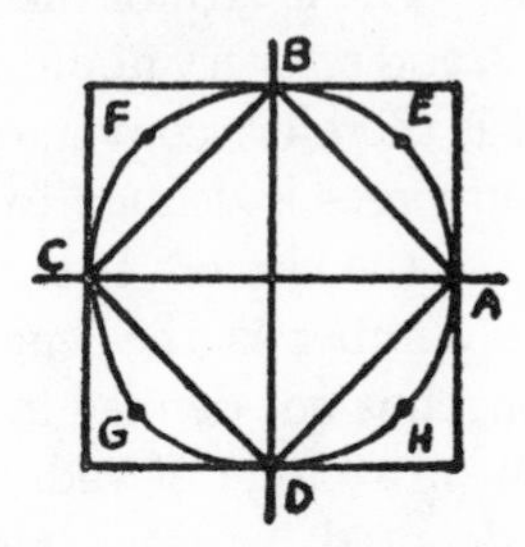

and the point B is (0, 1). The length of AB, i.e., the distance between these two points, is $\sqrt{2}$. The length of the perimeter of the square is therefore $4\sqrt{2}$.

Next we bisect each arc AB, BC, CD, DA, by points E, F, G, H. On joining up AE, EB, etc. we obtain a regular octagon inscribed in the circle. The length of this octagon can be found, though it does not have such a simple expression as the length of the square. The octagon is already a good deal closer to the circle than the square was.

And so we can proceed, at each stage inserting new points on the circle mid-way between the old ones. We obtain regular inscribed polygons with 16, 32, 64, $\cdots$ and generally 2^n sides. Let us denote the length of the perimeter of the polygon with 2^n sides by l_n.

Now I think it is clear from the figure that each l_n is greater than the one before. This simply follows from the fact that the sum of two sides of any triangle is greater than the third, since what we do in passing from l_n to l_{n+1} is to replace each side of the polygon of 2^n sides by two sides of that with 2^{n+1} sides. Hence the numbers l_n form a sequence, each term of which is greater than the term before it. On the other hand, the numbers l_n do not increase beyond all bounds. The perimeter of each of the inscribed polygons which we have used is less than that of the square of which A, B, C, D are the mid-points of the sides. This is fairly obvious from the figure, and anyhow mathematicians can easily prove it. The length of the perimeter of this square is equal to 8, and so every number l_n is less than 8.

The sequence l_n is therefore convergent, and the number to which it converges is defined to be the length of the circumference of the circle. It is not a question of proving that this number is the length. It is so by definition—the length is not defined in any other way.

Half the length of a circle of radius 1 is a number which is always denoted by the Greek letter π (pi). Thus the limit of the numbers l_n in the above construction is 2π. In any circle, the circumference is proportional to the radius, so that the circumference of a circle of radius r is equal to $2\pi r$.

It is obvious from the values of the perimeters of the two squares which we drew inside and outside the circle, that π lies somewhere between $2\sqrt{2}$ and 4.

The problem of area.

Possibly the idea of the area of a flat floor arose in connection with the problem of paving it with square tiles. One would want to know how many tiles were needed. Suppose that it is a rectangular floor, which could be exactly filled up with the tiles. If p of them go into it one way, and q the other way, then the total number required is $p \times q$. This number has nothing to do with the shape of the floor—two floors of different shapes with the same $p \times q$ are of equal importance from the tiling point of view. This number then deserves a special name, and it is called the *area*. The way in which we have obtained it clearly gives the rule for the area of a rectangle, area = length $\times$ breadth.

Even if the length and breadth are not exact multiples of a unit, as we have so far supposed, the same rule still gives a definite result, as long as the length and the breadth can be measured. This extends the definition of area to any rectangle.

We next require a rule for the area of a triangle. In Euclid's theory of triangles congruent triangles are regarded as being equal in all respects, so that, if they have areas, the areas must be equal. Now take any triangle ABC, and fit round it a rectangle as in the following figure.

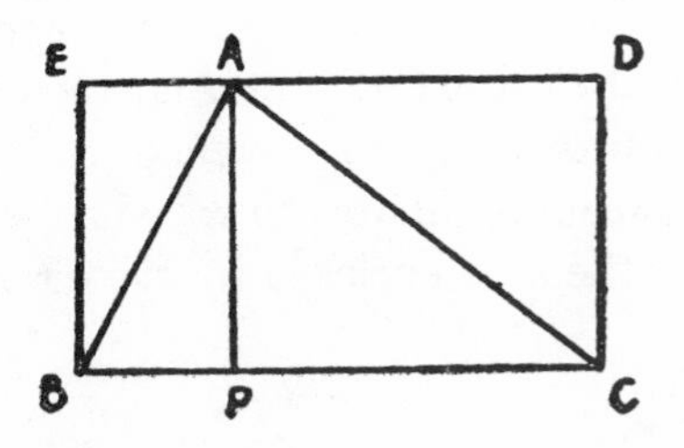

Here AP is perpendicular to BC. The triangles ABP and ABE are congruent and so of equal area. Similarly ACP and ACD are of equal area. Thus the triangle ABC must be half the area of the whole rectangle, i.e., half of $EB \times BC$, or half of $AP \times BC$. Hence we obtain the rule, area of triangle = half the base multiplied by the height.

Here we are using Euclidean geometry to give us hints as to how to proceed in certain cases, not as the logical base of our system. We must therefore really take the above rule as the definition of the area of a triangle. What the argument shows is that it is the only definition which is consistent with our ordinary geometrical ideas.

Later on we shall discuss the problem of area generally. It has been introduced here in the case of triangles because we want to discuss circles, and because it is a fairly simple step from the triangle to the circle.

Area of a circle.

Besides having a perimeter with a definite length, a circle encloses a region with a definite area. This can be seen by using the same construction as before. Take the regular polygon with 2^n sides inscribed in the circle, and join each vertex of it to the centre of the circle by a straight line. This divides the polygon into 2^n triangles. Now the area of each triangle is half the base multiplied by the height. Take as the base of each triangle the side which is one of the sides of the polygon. Then the sum of all the bases is the perimeter of the polygon, and so, when n is very large, is approximately equal to $2\pi r$ (if r is the radius). Also the height of each triangle is approximately r. Hence the area enclosed by the polygon approximates to πr^2 when n is large, and consequently the area enclosed by the circle is equal to πr^2.

The value of π.

Approximations to π are to be found in very ancient writings. In an Egyptian papyrus, written by Ahmes some time before 1700 B.C., and entitled "Directions for obtaining the Knowledge of all Dark things," the area of a circle is found by deducting from the diameter one ninth of its length and squaring the remainder (see F. Cajori, *A History of Mathematics*). If we take the radius to be 1, so that the diameter is 2 and the area π, this gives

$$\pi = \left(2 \times \frac{8}{9}\right)^2 = \left(\frac{16}{9}\right)^2 = \frac{256}{81}.$$

This is equal to $3.1604 \cdots$, a very fair approximation to π.

According to I Kings vii, 23, Hiram of Tyre made a molten sea, ten cubits from the one rim to the other; it was round all about, and a line of thirty cubits did compass it round about. According to this, the circumference of the "sea" was three times its diameter. If this were quite accurate, π would be equal to 3. No doubt Hiram, in his report to Solomon at any rate, ignored fractions of a cubit. If the molten sea was 9.6 cubits across, it would be about 30 cubits round, to the nearest cubit. If it was really ten cubits across, it must have been a good thirty-one cubits round.

The most celebrated approximation to π, $\frac{22}{7}$, is due to Archimedes. He actually showed that π lies between $3\frac{10}{71}$ and $3\frac{1}{7}$. A still better rational approximation to π is $\frac{355}{113}$. This is equal to $3.1415929\cdots$, and it agrees with π to six places of decimals.

The numerical value of π to fifteen decimal places is

$$\pi = 3.14159\ 26535\ 89793 \cdots.$$

It has been calculated to hundreds of decimal places, with what object it is difficult to say.

There are many striking formulae for π. In 1656, John Wallis, Savilian Professor of Geometry at Oxford, proved that

$$\pi = 2 \cdot \frac{2}{1} \cdot \frac{2}{3} \cdot \frac{4}{3} \cdot \frac{4}{5} \cdot \frac{6}{5} \cdot \frac{6}{7} \cdot \frac{8}{7} \cdot \frac{8}{9} \cdots$$

The expression on the right-hand side is an infinite product; that is, we are to multiply any finite number of factors, and take the limit of these "partial products" as the number of factors tends to infinity. This formula was the first in which π was expressed as the limit of a sequence of rational numbers. Perhaps the simplest such formula is the infinite series

$$\pi = 4\left(1 - \frac{1}{3} + \frac{1}{5} - \frac{1}{7} + \cdots\right)$$

published by Gregory in 1670. Another infinite series, this time for π^2, is

$$\pi^2 = 6\left(\frac{1}{1^2} + \frac{1}{2^2} + \frac{1}{3^2} + \cdots\right),$$

discovered independently by John Bernoulli and Euler.

Squaring the circle.

One type of problem which the ancient Greek geometers were fond of setting themselves was that of constructing a length with given properties. For example, they asked how to construct the side of a square which should be equal in area to a given triangle. "Construction" here had a special meaning. They were only allowed to use ruler and compasses; that is, a straight line through two given points could be constructed, and a circle with a given centre and radius of a given length could be constructed. Anything else must be made up by some combination of these processes. One might imagine the existence of other curves, but as they could not be so easily drawn in practice, it was not regarded as playing the game to use them in constructions.

The problem of the square equal in area to a given triangle was solved. Another problem which then suggested itself was that of constructing a square equal in area to a given circle; and of course it had to be done by Euclidean methods, i.e., with ruler and compasses only. The problem became known as that of squaring the circle. It was never solved, and we know now that it is *insoluble*. The proposed construction is an impossible one.

Let us see what this amounts to in terms of numbers. We may take the radius of the circle to be 1; its area is then π. The side of the proposed square would therefore have to be equal to $\sqrt{\pi}$.

It was proved by the German mathematician Lambert in 1761 that π is an irrational number. This is a very interesting discovery, but it does not prove that the problem of squaring the circle is impossible. Some irrational lengths, such as $\sqrt{2}$, can be constructed by Euclidean methods. But not every irrational length can be constructed by these methods. They lead to lengths of a special kind only, and it happens that $\sqrt{\pi}$ is not one of these. It was proved by another German mathematician, Lindemann, in 1882, that π is not merely irrational, but is what is called a *transcendental* number. This means that it is not a root of any algebraic equation with integer coefficients. Since every number which can be constructed by Euclidean methods is a root of such an equation, neither π nor $\sqrt{\pi}$ can be so constructed. Hence it is impossible to square the circle.

This of course does not mean that the proposed square does not exist. In fact we have shown above that it does. It merely means that it is impossible to construct it in the particular way used by the Greek geometers.

The three unsolved problems of antiquity.

There were three famous problems which were proposed by the ancient geometers, but which were never solved. One was that of squaring the circle. The second was that of trisecting a given angle, i.e., dividing a given angle into three equal parts. The third was that of duplicating a cube, i.e., to construct a cube which should have twice the volume of a given cube. All the constructions, of course, had to be done by Euclidean methods.

The ancients failed to solve these problems, not because they were not clever enough, but because the problems themselves were insoluble. This is true in each case for the same sort of reason, viz., that the solution would involve a kind of irrational number which cannot be constructed by Euclidean methods. Of course, particular angles can be trisected. The trisection of a right angle involves the construction of an angle of 30°, which can be done quite easily. But in general the problem is an impossible one.

It seems that the fame of these problems is worldwide, but the fact that they are insoluble is not so well known. There must be many people toiling pathetically on in garrets, trying to solve them still. Circle-squarers, and particularly angle-trisectors, still exist. They send me their solutions sometimes. Often the nature of the problem has been misunderstood, and it is thought that a good approximation to the solution is what is wanted. Usually the constructions proposed are so intricate that anyone might have gone wrong in the course of making them up. I must admit that I never try to hunt out the mistakes in these complicated figures. But I can assure the circle-squarers, angle-trisectors and cube-duplicators that there is general agreement among mathematicians that they have set themselves an impossible task.

The number e.

An expression which occurs in many mathematical formulae is the product of all integers up to and including a given integer. If the last integer is n, this product is called "factorial n," and is written $n!$ (In the older books it used to be written $\lfloor n$, but the line under the letter made this inconvenient to print.) Thus

$$n! = 1 \times 2 \times 3 \times \cdots \times n.$$

For example $1! = 1$, $2! = 2$, $3! = 6$ and $4! = 24$.

The number e is defined to be one plus the sum of the reciprocals of all the factorials. In symbols

$$e = 1 + \frac{1}{1!} + \frac{1}{2!} + \frac{1}{3!} + \cdots$$

or

$$e = 1 + \sum_{n=1}^{\infty} \frac{1}{n!}.$$

When we say "the sum of the reciprocals of all the factorials" we are of course speaking of the sum of an infinite series; e is not to be found just by addition, but it is the limit of a sequence in the sense explained in Chapter VIII. Since $n!$ increases very rapidly as n increases, $1/n!$ decreases very rapidly, and the reader can well imagine that the series just written down is convergent. This can be proved to be true, so that e is actually defined by the series. Numerical approximations to e can be found by taking the first few terms of the series and ignoring the remainder. An approximate value with fifteen decimal places is $e = 2.71828\ 18284\ 59045 \cdots$.

Why the number e is important in mathematics I hope to explain later. Here I shall only go as far as proving the following theorem: *e is an irrational number.*

In general, if a number is defined by a formula, for example as the sum of an infinite series, it is difficult to determine whether it is rational or irrational. The particular nature of the series which defines e makes it fairly easy to obtain the result in this case. The proof proceeds by *reductio ad absurdum.* That is, we assume for the sake of argument that the contrary of the proposed theorem is true, and show that the assumption leads to an absurdity or contradiction.

Suppose that e is a rational number; that is, that there are integers p and q such that $e = p/q$. Thus

$$\frac{p}{q} = 1 + \frac{1}{1!} + \frac{1}{2!} + \frac{1}{3!} + \cdots$$

Divide the series into two parts, the sum of the terms as far as $1/q!$ in the first part, and the remainder in the second part; thus

$$\frac{p}{q} = \left(1 + \frac{1}{1!} + \cdots + \frac{1}{q!}\right) + \left(\frac{1}{(q+1)!} + \frac{1}{(q+2)!} + \cdots\right).$$

Now multiply throughout by $q!$. In doing so, we treat the above expression as if it were a finite sum, and not an infinite series but it can be shown that in the case of a convergent series this is quite justifiable. The result is

$$p \times (q-1)! = (q! + \frac{q!}{1} + \frac{q!}{2} + \cdots + q + 1) + \frac{1}{q+1} + \frac{1}{(q+1)(q+2)} + \frac{1}{(q+1)(q+2)(q+3)} + \cdots.$$

The left-hand side is clearly an integer, and so is every term in the first bracket on the right-hand side. Hence the sum of the remaining terms is equal to the difference between two integers. This however is impossible; for the sum in question is clearly less than the corresponding sum in which each factor in each denominator is replaced by $q + 1$; and this is

$$\frac{1}{q+1} + \frac{1}{(q+1)^2} + \frac{1}{(q+1)^3} + \cdots$$

or

$$\frac{1}{q+1}\left\{1 + \frac{1}{(q+1)} + \frac{1}{(q+1)^2} + \cdots\right\}.$$

The series in the brace is simply an infinite geometrical progression, and its sum is

$$\frac{1}{1 - \frac{1}{q+1}} = \frac{q+1}{q}.$$

Hence the whole expression is equal to $1/q$. The corresponding sum in the previous expression has therefore been proved on the one hand to be equal to an integer, and on the other hand to be less than $1/q$; and this is a contradiction if q is greater than 1. The assumption that e is equal to a rational number p/q is therefore false.

This is one of the simplest cases in which a number defined by such a formula can be proved to be irrational. The proof that π is irrational is considerably more difficult.

A proof that the number π is transcendental, and consequently that it is impossible to "square the circle," is given by Hobson, *Plane Trigonometry*, 4th Edn., pages 305–311. The first step in this proof consists of proving that the number e is transcendental. Anyone who had merely read this chapter might not suspect that there was any connection between the number e and the number π. But there is a connection, expressed by the formula $e^{i\pi} = -1$. This formula anticipates several things that will be dealt with later in the book. The meaning of i is explained in Chapter X; $e^{i\pi}$ means the exponential series defined in Chapter XIII, with the variable replaced by $i\pi$. It is beyond the scope of this book to consider series involving i, but the theory of such series can be found in many other books.

Proofs that π and e are transcendental are also given by Hardy and Wright, *An Introduction to the Theory of Numbers*, Chapter XI.

All this seems very remote from the original geometrical figure which gave rise to the problem of squaring the circle. It is not surprising that more than two thousand years should have elapsed between the time when this problem was proposed, and the time when it was proved to be insoluble.

Proofs of the impossibility of trisecting the angle and duplicating the cube are given by Courant and Robbins, *What is Mathematics?* pp. 134–138.

Chapter X

THE SQUARE ROOT OF MINUS ONE

Quadratic equations.

As an introduction to this subject we shall first make a few remarks about quadratic equations. A quadratic equation is one which involves an unknown x, and also the square of x, but no cubes or any higher powers. Such for example are the equations

$$x^2 - 1 = 0, \tag{1}$$

$$x^2 - 4x + 3 = 0, \tag{2}$$

$$x^2 + 1 = 0, \tag{3}$$

$$x^2 - 4x + 5 = 0. \tag{4}$$

Let us consider what values of x satisfy these equations. In equation (1), on adding 1 to each side we obtain

$$x^2 = 1.$$

Hence x must be the square root of 1. There are two such square roots, 1 and -1. These therefore are the solutions, or roots of the equation, as they are called.

The second equation is a little more complicated; but if we recall from the formulae of pages 36–37 that

$$(x - 2)^2 = x^2 - 4x + 4$$

we see that the equation is equivalent to

$$(x - 2)^2 - 1 = 0.$$

Adding 1 to each side gives

$$(x - 2)^2 = 1.$$

Consequently $x - 2$ is either 1 or -1, and so x is either 3 or 1. These are the two roots of equation (2).

In each of these equations the roots happen to be integers; but it is easy to make up examples in which the roots are fractions or irrational numbers.

Now consider equation (3). Here the situation is quite different. Whatever x is, the square of x is a positive number (or at least 0), and consequently $x^2 + 1$ cannot be equal to 0. Hence this is an impossible equation; it has no solutions at all. Equation (4) is also an impossible one, since it is equivalent to

$$(x - 2)^2 + 1 = 0.$$

For the same reason as before, this has no solutions.

It might be thought that if some of these equations are soluble and others insoluble, we must just say so, and leave it at that. But it so happens that it is possible to construct an extended system of numbers, in terms of which problems of this kind are soluble. The situation is similar to that which we encountered before, in considering equations such as $2x = 1$. This is insoluble in terms of integers; but if we interpret it to mean a relation between the complex numbers (fractions) considered in Chapter IV, then it is soluble. As we have seen, the extension of the idea of numbers which this involves is very important.

So it is with the equations which we have encountered here. Some of them are insoluble if they are taken as referring to the numbers already known. But if we reinterpret them as referring to a certain system of complex numbers, then we find that they are soluble. As before, the new system of numbers which is introduced in this way turns out to be very interesting and important. These new numbers are of use not only in the theory of quadratic equations, which we have taken as a simple example, but in many other branches of mathematics as well.

Complex numbers.

Again we consider a system of complex numbers, each of which has two components, say x and y. We denote such a number by (x, y). We can begin by incorporating all the generalizations of numbers which have been made so far. Each of the two numbers x and y may be given any value, integral, fractional, or irrational.

The nature of this system of numbers now depends on the circumstances in which we regard two such numbers as equal, and on the laws of addition and multiplication which we impose upon them.

We say that two complex numbers (a, b) and (c, d) are equal if $a = c$ and $b = d$. Nothing could be simpler than this.

The law of addition is, in this case, the very obvious one

$$(a, b) + (c, d) = (a + c, b + d),$$

the two first components being added, and also the two second components. For subtraction we naturally also have

$$(a, b) - (c, d) = (a - c, b - d).$$

The law of multiplication is

$$(a, b) \times (c, d) = (ac - bd, ad + bc).$$

This is more complicated, and it is difficult to give any *a priori* reason for adopting this particular law. The reason will appear when we come to work out the properties of the resulting system of numbers.

Division may be defined as the inverse of multiplication; that is $(a, b) \div (c, d)$ must be a complex number (x, y) such that

$$(c, d) \times (x, y) \doteq (a, b).$$

In view of the law of multiplication, this is equivalent to

$$(cx - dy, cy + dx) = (a, b).$$

This means that the two relations

$$cx - dy = a, \qquad cy + dx = b,$$

must hold. It may be verified that this is true if

$$x = \frac{ac + bd}{c^2 + d^2}, \qquad y = \frac{bc - ad}{c^2 + d^2}.$$

The division law is therefore

$$\frac{(a, b)}{(c, d)} = \left(\frac{ac + bd}{c^2 + d^2}, \frac{bc - ad}{c^2 + d^2}\right).$$

Naturally it is assumed in this case that c and d are not both zero, or the right-hand side would be meaningless (both terms would be fractions with zero denominator).

This is our system, and we must now ask what it has to do with the ordinary system of numbers, and also what it has to do with the solution of quadratic equations. Consider in the first place the sub-class of these numbers in which the second component is zero, i.e., the class of numbers of the form $(a, 0)$. Two such numbers $(a, 0)$ and $(c, 0)$ are equal if $a = c$; and the laws of addition and multiplication are

$$(a, 0) + (c, 0) = (a + c, 0)$$

and

$$(a, 0) \times (c, 0) = (ac, 0).$$

These are got by putting $b = 0$ and $d = 0$ in the previous formulae.

Now these laws are of exactly the same form as the

laws of addition and multiplication of ordinary numbers. In fact the second component and the brackets can be mentally omitted. The result is that any mathematical operation in which ordinary numbers $a, c, \cdots$ appear, could be replaced by an exactly similar operation with numbers $(a, 0)$, $(c, 0)$, $\cdots$. Since it is the pattern, and not what makes the pattern, that matters, we can to all intents and purposes identify $(a, 0)$ with a.

Next consider complex numbers in which the first component is zero. Addition for these goes as before; the rule is

$$(0, b) + (0, d) = (0, b + d).$$

But for multiplication we have

$$(0, b) \times (0, d) = (-bd, 0).$$

In the product, it is the second component which is zero. According to what we have said about such numbers, this is to be identified with the ordinary number $-bd$. As particular cases we have

$$(0, b) \times (0, b) = (-b^2, 0)$$

and

$$(0, 1) \times (0, 1) = (-1, 0).$$

These are remarkably different from the ordinary formulae for squares of numbers. The square of any number in this system is in fact negative, not positive.

We see at once that there is no difficulty whatever about solving the equation $x^2 + 1 = 0$, if we interpret x as a complex number (a, b), 1 as the complex number $(1, 0)$, and 0 as the complex number $(0, 0)$. We want

$$(a, b)^2 = (0, 0) - (1, 0) = (-1, 0),$$

and a solution is given by the last formula above, viz., $a = 0, b = 1$. Hence $x = (0, 1)$. Actually there is also a second solution, $x = (0, -1)$.

In this system, then, there are two square roots of -1 [really, of course, of $(-1, 0)$]. The former $(0, 1)$ is denoted by i, and the latter $(0, -1)$ by $-i$.

Now consider any complex number (x, y). By the laws of addition and multiplication

$$(x, y) = (x, 0) + (0, y) = (x, 0) + (0, 1) \times (y, 0).$$

This formula makes it possible to dispense with the rather cumbrous bracket notation altogether. We write simply x instead of $(x, 0)$, and i instead of $(0, 1)$. Then

$$(x, y) = x + iy.$$

We can now operate with these numbers in the same way as with ordinary numbers, provided that, if i^2 occurs, we replace it by -1. For example, the law of multiplication reappears at once; we have

$$\begin{aligned}(a, b) \times (c, d) &= (a + ib) \times (c + id) \\ &= ac + aid + ibc + i^2bd \\ &= ac - bd + i(ad + bc) \\ &= (ac - bd, ad + bc).\end{aligned}$$

Historically, of course, all this was done first the other way round. Problems occurred in which the formulae required that -1 should have a square root. Very well then, mathematicians said, in spite of everything, let us pretend that it *has* a square root. It might have been expected that this would lead to one absurdity after another, but in fact it did not. If we denote the fictitious square root of -1 by i, and replace i^2 wherever it occurs by -1, we get a reasonable system of algebra; in fact of course just the algebra which we have made up out of the complex numbers with two components.

It is of some importance to present the theory in the way which has been taken here, because even to-day many people suppose that there is something mysterious about the square root of minus one. This is entirely due to the historical development of the subject. I do not see how anyone can find anything sinister or contradictory about the number-pair (0, 1).

If one overheard an argument between two people, the first of whom said that -1 had a square root, while the other maintained that it had not, one would probably have to say, "You are both right, but you are talking about different things. The old original -1 has no square root, in the system to which it naturally belongs, but if you attach the same name to the number-pair $(-1, 0)$, then it certainly has a square root, and in fact it has got two of them. They are (0, 1) and $(0, -1)$, familiarly known as i and $-i$."

I met a man recently who told me that, so far from believing in the square root of minus one, he did not even believe in minus one. This is at any rate a consistent attitude.

There are certainly many people who regard $\sqrt{2}$ as something perfectly obvious, but jib at $\sqrt{(-1)}$. This is because they think they can visualize the former as something in physical space, but not the latter. Actually $\sqrt{(-1)}$ is a much simpler concept.

The air of mystery about the subject has been preserved by some unfortunate names which have been used in it. Numbers of the form $(x, 0)$, regarded as equivalent to the ordinary x, have been called "real," while numbers of the form $(0, y)$, or iy, have been called "imaginary." Thus $x + iy$ is the sum of "real" and "imaginary" parts. The ordinary meaning of these words does not correspond to anything in the nature of the numbers so described.

Anyone who had an insuperable objection to the use of the complex numbers of this system could actually prove without them everything that can be proved with them. For every relation between complex numbers is simply equivalent to two relations between real numbers. Suppose for example we are confronted with the relation $z = w^2$, where z and w are complex numbers, say $z = (x, y) = x + iy$, and $w = (u, v) = u + iv$. Now by the law of multiplication

$$w^2 = (u, v) \times (u, v) = (u^2 - v^2, 2uv).$$

It follows that $x = u^2 - v^2$ and $y = 2uv$. These two relations between "real" numbers are equivalent to the relation $z = w^2$ between complex numbers.

This process is known as "equating real and imaginary parts." It is sometimes necessary, of course, but often it would not only double the labour of calculation, but would obscure the whole point of formulae containing these numbers.

The Argand diagram.

Complex numbers are often represented by points in a plane, the number (x, y) corresponding to the point (x, y) in two-dimensional Cartesian geometry. This representation is known as the Argand diagram.

In this way of looking at it, addition and multiplication have fairly simple geometrical meanings. The sum $z + z'$ corresponds to the fourth vertex of the parallelogram, three of whose vertices are the points O, z and z'. Multiplication is not quite so simple. Suppose that the point z is at a distance r from O, and that the line joining z to O makes an angle A with the axis of x. Similarly let z' be at distance r' from O, and let the line joining z' to O make an angle B with the axis of x. Then the product zz' corresponds to a point at a distance rr' from O, and such that the line joining it to O makes an angle $A + B$ with the axis of x. The reader should draw a figure to show this. The proof requires a little trigonometry, so that we cannot give it here.

Solution of algebraic equations in terms of complex numbers.

The two quadratic equations which, on page 117, we had to leave as insoluble were $x^2 + 1 = 0$ and $x^2 - 4x + 5 = 0$. If we interpret these in terms of complex numbers, the roots of the former are i and $-i$, as we have just seen. The latter is equivalent to

$$(x - 2)^2 = -1.$$

Hence $x - 2 = i$ or $-i$, and the roots are $x = 2 + i$ and $x = 2 - i$.

It can easily be shown that any quadratic equation can be solved, if it is interpreted in this way.

It might be supposed that to solve equations of higher order we should require complex numbers of still more complicated kinds; for example that a cubic, or equation involving x^3, would involve a second set, a quartic, or equation involving x^4, a third set, and so on. This is a mathematician's nightmare, which turns out to have no foundation in fact. An important theorem, sometimes called the fundamental theorem of algebra, says that any algebraic equation has a solution, if it is interpreted in terms of the system of complex numbers already introduced. The proof of this theorem is difficult, and it is impossible to give any idea of it here. The effect of it (and other similar theorems) is that this system of complex numbers is, in a whole class of problems, all that is wanted. We have to make this generalization of the idea of number, but, once we have got it, we do not need anything else.

There are, of course, many other systems of complex numbers. In one such system, called quaternions, each number has four components. Another system has recently been used by Eddington in developing his theory of the universe. In Eddington's system there are no fewer than sixteen square roots of minus one. This system was used by Eddington, not at all for fun, but because it appeared to be the best method of representing certain aspects of the physical world.

Chapter XI

TRIGONOMETRY

Trigonometry is mostly about measuring triangles. It is an art whereby, if certain parts of a triangle are known, the others can be calculated. A triangle has three sides and three angles. Let us denote the sides by a, b, and c, and the angles opposite to them by A, B, and C.

If the lengths of all the sides are known, then the shape of the triangle is fixed, and the angles can be determined. Again, suppose that two sides, say a and

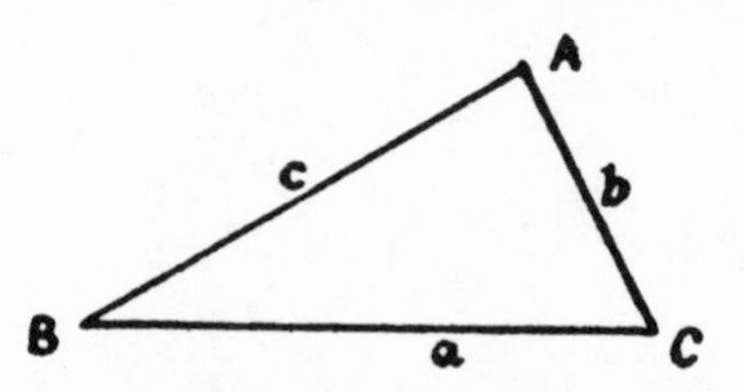

b, and the angle C included between them, are given. Then again the triangle is fixed, and the third side c and the other two angles A and B can be found. Still another case is that in which one side, say a, and the two angles B and C at its ends are given. Then again the triangle is fixed, and the other two sides b and c, and the third angle A can be found. This process of finding the remaining parts from those which are known is called solving the triangle.

Trigonometry has applications well known to surveyors and men who map out the world. Suppose for example that the line BC is measured out on land, and that A is a point at the same level which we can see in the distance. Let us stand at B and measure the angle there, and then at C and measure the angle there. Then the triangle can be solved, and so the distance of A from either B or C can be found.

In another case, A is a distant mountain top. Suppose that the horizontal distance a of the mountain from an observer at B is known, and the angle B through

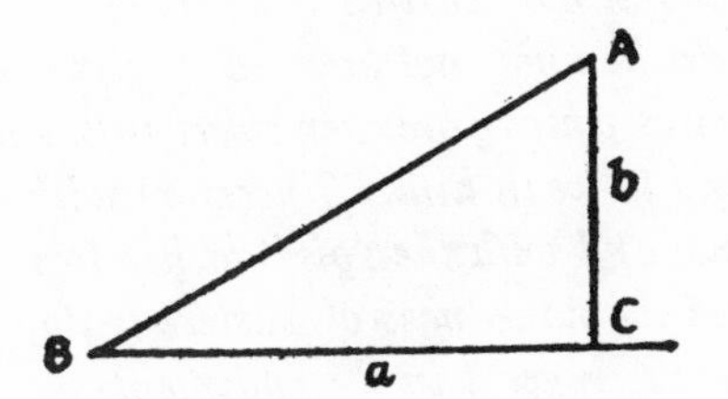

which we have to raise our eyes to look at A can be measured. The angle at C is also known (it is a right angle). Hence the triangle can be solved, and the height b of the mountain can be found.

If we can suppose that the familiar geometry of objects on earth is true also for those which we see out in space, these methods can also be used to find the distance of the heavenly bodies. BC can be a base-line measured on the earth, or even the distance between a point on the earth at two different places in its orbit. The distance of a heavenly body A can then be determined by observation of the angles at B and C.

Reader, do not despise these studies. They solaced the youth of Wordsworth's "Wanderer."*

Lore of a different kind
The annual savings of a toilsome life
His stepfather supplied; books that explain

* The Excursion, Bk I.

The purer elements of truth involved
In lines and numbers, and by charm severe
(Especially perceived where nature droops
And feeling is suppress'd) preserve the mind
Busy in solitude and poverty.
These occupations oftentimes deceived
The listless hours, while in the hollow vale
Hollow and green, he lay on the green turf
In pensive idleness. What could he do
With blind endeavours, in that lonesome life,
Thus thirsting daily? Yet still uppermost
Nature was at his heart as if he felt—
Though yet he knew not how—a wasting power
In all things which from her sweet influence
Might tend to wean him. Therefore with her lines,
Her forms, and with the spirit of her forms,
He clothed the nakedness of austere truth,
While yet he linger'd in the rudiments
Of science, and among her simplest laws,
His triangles—they were the stars of heaven,
The silent stars! Oft did he take delight
To measure th' altitude of some tall crag
That is the eagle's birthplace, or some peak
Familiar with forgotten years, that shews
Inscribed, as with the silence of the thought,
Upon its bleak and visionary sides
The history of many a winter storm,
Or obscure records of the path of fire.

How to measure an angle.

We have spoken above of measuring the sides and angles of a triangle. According to the system of geometry which we have explained, the vertices of a triangle are points in the Cartesian plane defined by co-ordinates (x, y), and the distance between two of them (x, y) and (x', y') is defined by the formula $\sqrt{\{(x - x')^2 + (y - y')^2\}}$ given before. We have not yet said what is meant by measuring an angle, and this problem must now be faced.

Suppose that the vertex of the angle is made the centre of a circle of radius 1. Then a measure of the size of the angle is the length of the part of the circle lying between the arms of the angle. That this length exists can be proved in the same way as was used to prove that the whole circle has a length.

The measure assigned to an angle in this way is called its circular measure. The circular measure of a complete turn (e.g., facing north round to facing north again) is the whole circumference of the circle, and so it is equal to 2π. The circular measure of a right angle is a quarter of that of a complete turn, and so is $\frac{1}{2}\pi$. At any rate, it is fairly obvious that this is the measure of the angle in the Cartesian plane between any of the axes of co-ordinates and the next one round beyond it. We defined perpendicular lines in general as two lines $lx + my = 0$ and $l'x + m'y = 0$ such that $ll' + mm' = 0$, and strictly speaking we ought to prove that the circular measure of the angle between any two such lines is $\frac{1}{2}\pi$. This is not difficult to do, and we shall assume here that all such results, involving the direction in which the angle sticks out, are correct.

There are a few other angles of which the circular measure can easily be written down. It is known that the angles of any triangle together make up two right angles, i.e., an angle of circular measure π. The angle of an equilateral triangle must be one-third of this, and so its circular measure is $\frac{1}{3}\pi$.

The unit of circular measure is known as a radian; that is, a radian is an angle such that the length of the arc of a circle of radius 1 drawn across it is also equal to 1.

The traditional way of measuring an angle is rather different from this. It originated with the Babylonians, who counted in the scale of sixty, and consequently found one-sixtieth to be a specially convenient fraction. According to this system, the whole circle is divided up into 360 equal parts, and the angle subtended by each part at the centre is called a degree. One sixtieth part of a degree is called a minute, and one sixtieth part of a minute is called a second (i.e., second-sixtieth). Thus the circular measure of a degree is $\frac{\pi}{180}$, that of a minute is $\frac{\pi}{10800}$, and that of a second is $\frac{\pi}{64800}$. An angle of 7 degrees, 6 minutes and 5 seconds, for example, is written $7^\circ\ 6'\ 5''$.

The difficulty about this is, how to divide a circle into 360 equal parts, if we have not already defined the measure of an angle in some other way? It can be done of course, by methods of repeated bisection and approximation, but this comes to much the same thing as the definition by means of circular measure already used.

Apparently the construction of a single degree by Euclidean methods (ruler and compasses) is not possible. The problem is the same as that of constructing a regular polygon of 360 sides, since there are 360 degrees in a complete turn. Since there is a simple Euclidean method of bisecting any angle, it would be sufficient to begin by constructing a regular polygon of 45 sides; for we could then use the bisection method three times to construct in turn polygons of 90, 180 and 360 sides. But the regular polygon of 45 sides is not one of those which can be constructed.

Euclid gave constructions for regular polygons of 3, 4, 5, 6, 8, 10, 12 and 15 sides. The construction is also possible for any number of sides which can be formed from these by multiplication by any power of 2. There are also some other regular polygons which can be constructed, the most interesting being that of 17 sides.

It was proved by Gauss that the construction is possible for any polygon with $2^{2k} + 1$ sides, where k is any integer, provided that this is a prime number. The numbers 3, 5, 17, 257 and 66537 correspond to $k = 0$, 1, 2, 3, 4, and all these are prime numbers, so that all such regular polygons can (in theory) be constructed. Several mathematicians have actually worked out the construction for 17 sides, and this is complicated enough. The case of 257 sides has been worked out.

There is a story that a student at Göttingen once asked his professor for a subject on which to write a thesis. The professor, who thought that in any case he would do no good, told him (as a joke) to construct the regular polygon of 66537 sides. The student went away. Many years afterwards, an old man tottered into the university, carrying a heavy trunk. He had come back with the construction of his polygon.

The trigonometrical ratios.

Actually it is quite possible to measure angles without using the lengths of circles, or any sophisticated ideas of that kind. For example, instead of drawing a circle across our angle, and measuring an arc of that, we could

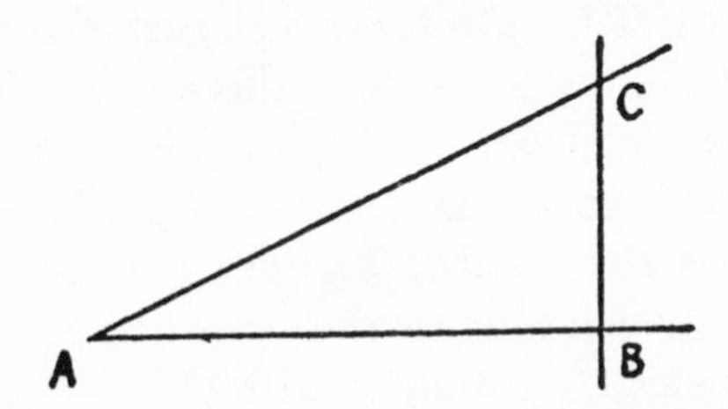

erect a perpendicular to one of the arms, say at unit distance along it from the vertex. The angle can then be measured by the length BC cut off on this perpendicular by the other arm of the angle. This way of measuring an angle involves nothing more than the measurement of distances between points.

We might start by taking a point B on one arm of the angle at a distance from A different from 1. In virtue of the properties of similar triangles, the ratio of BC to AB, i.e., the fraction (length BC) $\div$ (length AB), is the same, whatever point B we start from. The same measurement of the angle is therefore given by this ratio, which is independent of the actual lengths of AB and BC. For this reason, this measure of the angle is called a trigonometrical ratio.

This particular measure of the angle A is called the tangent of the angle, and is written for short as $\tan A$. This name arose as follows. It was customary to consider an angle as part of the following figure.

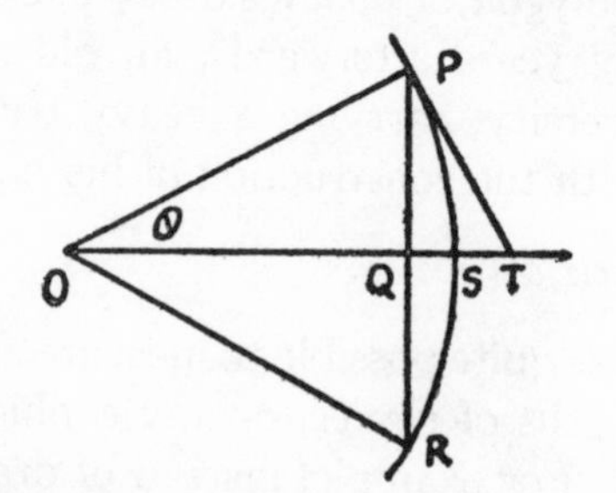

Here POR is twice the angle to be considered, and it is bisected by OQ, so that we really consider the angle POQ. The circular measure of this angle is denoted by the Greek letter θ (theta = th). PR is an arc of a circle with centre O and radius 1, and OQ produced meets it in S. PT is the tangent to the circle at P, i.e., the straight line which touches the circle at P. It is known that this makes a right angle with OP.

The tangent of the angle θ, $\tan \theta$, is then equal to the length of the tangent PT, since it is equal to PT/OP, and OP is 1.

As examples, we may give the following values of the tangents of angles:

$$\tan 30^\circ = \tan \tfrac{1}{6}\pi = 1/\sqrt{3},$$

$$\tan 45^\circ = \tan \tfrac{1}{4}\pi = 1,$$

$$\tan 60^\circ = \tan \tfrac{1}{3}\pi = \sqrt{3}.$$

A right-angle, $90°$ or $\frac{1}{2}\pi$ radians, has no finite value for its tangent. If we think of P, in the above figure, travelling round the circle until it is vertically above O, T travels farther and farther away to the right. Ultimately the tangent at P is parallel to OQ, so that they never meet, and the point T does not exist. Since the length PT increases beyond all bounds in this process, we express the result in the picturesque phrase "the tangent of a right angle is infinite."

Another way of measuring the angle θ is by means of the length of the line PQ. This is called the sine of the angle, and is written $\sin\theta$ (but sin is pronounced sine). The sine is another trigonometrical ratio, and is an alternative way of measuring an angle. Some examples of the sines of angles are:

$$\sin 30° = \frac{1}{2},$$

$$\sin 45° = \frac{1}{\sqrt{2}},$$

$$\sin 60° = \frac{\sqrt{3}}{2},$$

$$\sin 90° = 1.$$

Still another way of measuring the angle θ is by means of the length OT. This is called the secant, and is written $\sec\theta$.

The ratios $1/\tan\theta$, $1/\sec\theta$, and $1/\sin\theta$ are called the cotangent, cosine and cosecant of the angle, and are written $\cot\theta$, $\cos\theta$, and $\operatorname{cosec}\theta$.

Naturally all these ways of measuring the angle θ are related. If ABC is a triangle with a right angle at B, and the angle at A is θ, then $\sin\theta = BC/AC$, $\cos\theta = AB/AC$, and $\tan\theta = BC/AB$. Consequently $\tan\theta = \sin\theta/\cos\theta$. Another important relation is

$$\sin^2\theta + \cos^2\theta = 1.$$

Here $\sin^2 \theta$ means $\sin \theta \times \sin \theta$, i.e., the square of $\sin \theta$ ($\sin \theta^2$ would mean the sine of the square of θ, which would be quite different): and similarly $\cos^2 \theta$ means $\cos \theta \times \cos \theta$. The relation is equivalent to

$$\frac{BC^2}{AC^2} + \frac{AB^2}{AC^2} = 1,$$

or

$$BC^2 + AB^2 = AC^2.$$

This is true by Pythagoras' theorem (or of course in Cartesian geometry by the definition of distance).

Trigonometry books are full of formulae relating the sides and angles of a triangle. Here we can mention two of the simplest only. Suppose that we have a triangle ABC. Let the lengths of the sides opposite to the vertices A, B, C, be a, b, and c respectively, and suppose the three angles measured by their sines, $\sin A$, $\sin B$, and $\sin C$. Then the sine of the three angles are proportional to the length of the opposite sides; in formulae

$$\frac{\sin A}{a} = \frac{\sin B}{b} = \frac{\sin C}{c}.$$

This is a pleasing and elegant result which should appeal to anyone with an eye for a nice formula. Apart from this, it happens to be useful. Suppose that we know the length of the side a of the triangle, and the measures of the angles B and C. Then since the angles of a triangle add up to two right angles, we can at once determine the angle A. We can therefore find the values of $\sin A$, $\sin B$, and $\sin C$; and then the lengths b and c of the other two sides can be found from the formula. In fact they are

$$b = a\frac{\sin B}{\sin A}, \qquad c = a\frac{\sin C}{\sin A}.$$

This is just what we need to determine the distance of

an inaccessible point, if we can measure out a base line, and get the bearings of the point from each end of the base line. As another example of the solution of a triangle, suppose that we are given the lengths of the three sides, a, b, and c. Then it is required to find the angles, A, B, and C. The formula actually determines the cosines of these angles. It is

$$\cos A = \frac{b^2 + c^2 - a^2}{2bc},$$

with similar formulae for cos B and cos C. Having determined the cosines, the angles can be found, e.g., in degrees, minutes and seconds.

As another example of the use of trigonometry, suppose that we wish to measure the altitude of a tall crag, whose altitude and distance from us are both unknown.

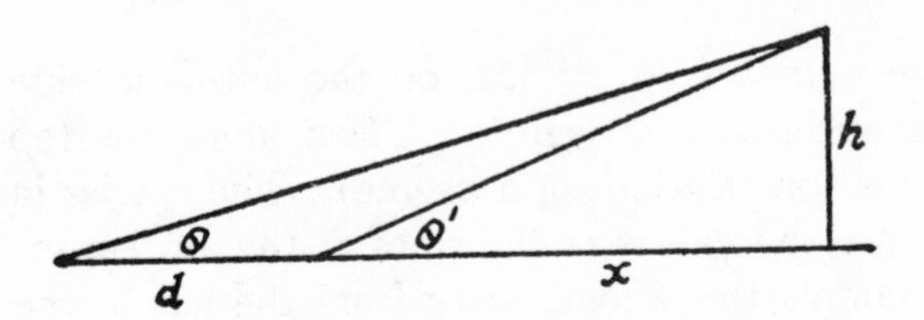

We can measure its elevation θ from the point where we stand. Now walk towards it a distance d. The elevation will now be greater; call it θ'. We are now at an unknown distance x from a point on the same level as the observer, vertically below the top. However, the various parts of the figure are connected by the formulae

$$\cot \theta = \frac{d + x}{h} \qquad \cot \theta' = \frac{x}{h}.$$

On subtracting we obtain

$$\frac{d}{h} = \cot \theta - \cot \theta'.$$

Since d, θ and θ' are known, and the cotangents can be calculated, the value of h can be found from this formula.

Trigonometrical tables.

In the previous section, we have made statements such as "An angle A is known; therefore its sine can be determined." How is this to be done? The answer usually is, by getting someone else to do the work. In other words, we quote the value printed in a table.

Here is a small specimen taken out of a table of sines.

	0′	10′	20′	30′	40′	50′
30°	.50000	.50252	.50503	.50754	.51004	.51254
31°	.51504	.51753	.52002	.52250	.52498	.52745
32°	.52992	.53238	.53484	.53730	.53975	.54220

The numbers 30, 31, 32, on the left-hand side are angles measured in degrees. Those along the top are minutes (one sixtieth of a degree). The entries in the table are the values of the sines of the angles, so that for example the second entry from the left in the top row, .50252, is the value of the sine of the angle of 30° 10′ (thirty degrees and ten minutes). These values are of course not exact. They are the best we can do with only five decimal places at our disposal; that is, we write down the rational number with denominator 100000 which is nearest to the actual value of the sine. The complete table runs from 0° to 90°. It is quite a simple and crude table. Many more accurate ones exist, both as regards the number of decimal places used and the closeness between the entries, which here is 10′.

How did the man who made the table calculate these values? The methods are too complicated to describe here in detail. For some angles, the exact value of the sine is known. For example, sin 30° is exactly $\frac{1}{2}$, and this appears as the first entry, .50000, in the above table. For most angles there is no exact formula, but there are approximate formulae, and these are used to fit in the values of the sines of such angles.

Similar tables are published giving the values of the tangents of angles. It is such a table that would be used in the problem of the crag referred to above, since $\cot \theta = 1/\tan \theta$, so that the cotangent can easily be deduced from the tangent. We also often require the values of cosines, but these can be deduced from a table of sines, since the cosine of an angle is equal to the sine of 90° minus the angle. The reader should be able to verify this without difficulty from a figure.

The addition formulae.

I will record here, without attempting to prove them, some formulae which play a very important part in trigonometry. Suppose that we have any two angles, which we may denote by A and B. Then the sine or cosine of $A + B$ can be expressed in terms of the sines and cosines of the separate angles A and B. The formulae are

$$\sin (A + B) = \sin A \cos B + \cos A \sin B,$$

and

$$\cos (A + B) = \cos A \cos B - \sin A \sin B.$$

These are called the addition formulae for the sine and cosine. Proofs of them are to be found in every book on trigonometry.

As a particular case we might take the two angles equal, say $B = A$. The formulae in this case reduce to

$$\sin 2A = 2 \sin A \cos A,$$

and

$$\cos 2A = \cos^2 A - \sin^2 A.$$

We can thus find the sine or cosine of twice an angle, if

we know the sine or cosine of the angle. A more interesting step is usually in the other direction. In view of the relation $\sin^2 A + \cos^2 A = 1$, the formula for $\cos 2A$ can be written in the form $\cos 2A = 2 \cos^2 A - 1$, or in the form $\cos 2A = 1 - 2 \sin^2 A$. These formulae enable us to find the sine or cosine of half an angle from the cosine of the angle. For example from $\cos 45° = 1/\sqrt{2}$ we deduce that $\sin 22\frac{1}{2}° = \sqrt{\frac{1}{2} - \frac{1}{2\sqrt{2}}}$

Demoivre's theorem.

There is a remarkable connection between the formulae of trigonometry and the formulae in the theory of the complex numbers considered in the last chapter. The original result in this order of ideas is called after its discoverer Abraham de Moivre or Demoivre (1667–1754), a mathematician of French birth who lived in London. It is:

$$(\cos \theta + i \sin \theta)^n = \cos n\theta + i \sin n\theta,$$

where i is the "square root of minus one." It is not difficult to derive this from the addition formulae, at any rate when n is a positive integer. For example, when n is 2, the left-hand side is

$$(\cos \theta + i \sin \theta) \times (\cos \theta + i \sin \theta) = \cos^2 \theta + 2i \sin \theta \cos \theta + (i \sin \theta)^2.$$

According to the rule for interpreting i^2, this is equal to

$$\cos^2 \theta - \sin^2 \theta + 2i \sin \theta \cos \theta,$$

and, by the above formulae for the sine and cosine of twice an angle, this is $\cos 2\theta + i \sin 2\theta$, which is Demoivre's theorem.

Spherical trigonometry.

There is another sort of trigonometry in which we measure triangles, not in a plane, but on the surface of a sphere. On a sphere there are of course no straight lines, and the simplest curves on a sphere are circles. Any plane intersects the sphere in a circle, and a plane through the centre of the sphere intersects the sphere in the greatest possible circle. Such a circle is called a great circle, and great circles play the same part on a sphere as straight lines do on a plane. The earth is roughly spherical; on the earth, the equator and the meridians are examples of great circles. The situation on a sphere is of course not quite the same as on a plane, because two great circles intersect in two points (e.g., two meridians intersect at the north pole and also at the south pole), whereas two straight lines intersect in at most one point. Still, it is usually possible to concentrate our attention on one of these intersections.

A spherical triangle is a figure on a sphere bounded by arcs of three great circles. A spherical triangle has three sides and three angles, just as a plane triangle has; but the sides are measured, not by their lengths, but by the angles which they subtend at the centre of the sphere. So in spherical trigonometry the angles are angles, and the sides are angles too. If we wish to survey the surface of the earth, taking into account its curvature, it is spherical trigonometry which we have to use. The problems of spherical trigonometry, solution of triangles and so forth, are similar to those of plane trigonometry, but naturally the formulae are more complicated. I will quote only one of them. Suppose that we have a spherical triangle, whose angles are A, B, and C, and whose sides are a, b, and c (i.e., the side opposite to the angle A subtends an angle a at the centre of the sphere). Then

$$\frac{\sin A}{\sin a} = \frac{\sin B}{\sin b} = \frac{\sin C}{\sin c}.$$

This is an elegant formula, which obviously has some connection with the corresponding formula of plane trigonometry.

Chapter XII

FUNCTIONS

In ordinary life we are familiar with many things, the measure of which depends on some other thing, which can also be measured. The temperature at noon depends on the season of the year. The temperature at a particular place depends on the time of day as well as on the season. The position of the hands of the clock depends on the time of day. The atmospheric pressure indicated by the barometer depends on the height above sea-level. The force exerted by the sun on a planet depends on the distance between the sun and the planet. The income-tax which I pay depends upon my income.

In all such cases, that which depends is called a function of that on which it depends; the temperature is a function of the season, and so on. Scientists are most interested in functions which can be measured with some accuracy, and so can be represented by numbers and formulae. In mathematics, our interests are mainly in the numbers and formulae themselves. We are interested in functions, but they are idealized functions, in which the rules of dependence are those which we make up ourselves. Of course we do this in many cases according to suggestions offered to us by Nature.

Turning back over the pages of this book, the reader will find several examples of such functions. The area of a circle is a function of the radius. If the radius is r, and the area A, then $A = \pi r^2$. Also the length of the circumference is a function of the radius; if it is L, then $L = 2\pi r$. In trigonometry there are several interesting functions. If θ is the circular measure of an angle, the sine, cosine, tangent etc., of the angle, are functions of θ. They are denoted by the formulae $\sin\theta$, $\cos\theta$, $\tan\theta$.

Most of the functions of real life are too erratic, and depend on too many unknown factors, to be comparable with mathematical functions, or, as we say, to be reduced to a formula. There is no formula for the noon-day temperature. There is a formula for the angle turned through by the minute hand of the clock. It is $\theta = 6t$, where θ is measured in degrees, t is the time in minutes, and θ is measured from its position at the zero-hour from which the time is measured. This formula, however, really applies only to an ideal clock, with which the observed clock is roughly comparable.

The income-tax payable on a given income is determined by an exact formula. In this case the formula is fixed first by act of parliament, and the physical reality follows the formula. It can do so exactly, because it applies to finite sets of discrete objects (sums of money), the amount of which is an integer, and so not subject to fractional errors.

In the theory of functions, the number which depends on another is called the dependent variable, and that on which it depends is called the independent variable. The word variable is used because of the suggestion of such examples as the hands of a clock, which vary as time goes on. We need not necessarily think in this way. For example, we can just think of all possible radii of circles, and of each radius having attached to it another number, the area of the corresponding circle. But it is more exciting to think of the radius as starting from nothing and growing steadily larger. The circle then expands steadily as the radius grows, and its area grows too. The radius is then thought of as the independent variable, and the area as the dependent variable.

The word variable suggests its opposite, a constant. This is used in slightly different ways. We call π a constant, or sometimes an absolute constant. Its value is determined once and for all by the primary assumptions of mathematics. Other constants depend on comparatively local circumstances. In some cases, a variable can be a constant, if it just happens not to vary. Consider for example the income-tax £y, payable on an income of £x per annum. In accordance with the general scheme, we call x the independent variable, and the tax y the dependent variable. But actually the tax is 0 over a certain range of small values of x. Its value is thus constant over this range. A function can of course be a constant for all values of the independent variable; thus the ratio of the circumference of a circle to the radius is constant for all values of the radius.

Functions defined by formulae.

In mathematics a function is usually defined by means of a formula. We have already had some examples of this, and it is clear that we can construct any number of others. For example, if $y = 2x + 1$, or $y = x^2 + 2x + 3$, or $y = x^4 + x^2 + 1$, then y is a function of x. To every value of x corresponds a value of y, which can be calculated from the formula. In other cases, such as $y = \sqrt{(x - 1)}$, y is defined only for certain values of x, in this case for $x \geqslant 1$ (that is, if we are thinking only of "real" numbers); for if $x < 1$, there is no "real" square root of $x - 1$.

The usual notation for a function in such cases is $y = f(x)$. The "f" stands for "function." One can think of the $f(\ )$ as a machine into which the value of x is to be fed, and from which will then emerge the corresponding value of y. We can also write $f(x)$

$= 2x + 1$, $f(x) = x^2 + 2x + 3$, etc., if we wish to say which particular function we are considering. Naturally, any other letter could be used equally well instead of f. The Greek ϕ is used almost as much—$y = \phi(x)$.

Sometimes more than one formula is used to define a function. We might say for example "let $y = x$ if $0 \leqslant x \leqslant 1$, and let $y = x^2$ if $x > 1$." These rules can be thought of as defining a single function $y = f(x)$, though the formulae are different in different intervals. Such functions are familiar to taxpayers, since the formula which determines the tax is different for different ranges of income.

In very simple cases it is possible to define the function merely by specifying its value in certain intervals. We might say, for example "let $y = 1$ for $0 \leqslant x \leqslant 1$; let $y = 2$ if $1 < x \leqslant 2$, and let $y = 3$ if $x > 2$." Here x does not appear in the formulae, but only in the rules about the ranges to which they are to apply. Such a function is constant throughout an interval, and then steps up (or down) to another value in another interval. It is often called a step-function. A step-function has recently been proposed as the rule by which a man whose income is x shall contribute to the university education of his children. Naturally the steps go up in this case, but the result would be that a father with an income just above a "step" would be in a worse position financially than one with an income just below it. It is strange that the wise men who make our laws are unable to think of functions which vary in more subtle ways than step-functions.

Functions represented by graphs.

Perhaps the most familiar example of a graph in ordinary life is the temperature chart of a patient in a fever. At certain times of the day the nurse takes your temperature. The time at which it is taken, say t, is represented by t units of measure along the chart from the left-hand side. There is a line across the chart representing the normal temperature. If the temperature of the patient is, say, 2 degrees above normal, this fact is represented by a dot on the chart two units above the "normal" line. If the temperature of the patient were subnormal, say by 1 degree, this would be represented by a dot one unit of distance below the normal line.

As time goes on, a succession of dots is thus made on the chart. They are usually joined up rather roughly by lines, to suggest what must have happened in between the times at which the temperature was taken.

A diagram of this kind, in which the relation between an independent variable (the time t in the above case) and a dependent variable which is a function of it (the temperature in the above case) is called a graph. It does indeed represent graphically the behaviour of the function.

A similar diagram can be used to represent the functions defined by mathematical formulae. To take a very simple case, suppose that y is defined as a function of x by the formula $y = x + 1$. Take a piece of paper,

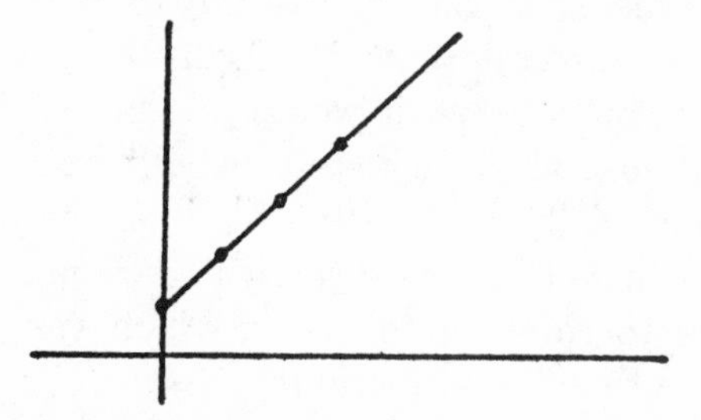

and draw on it two lines at right angles, one across and the other up-and-down the paper; or preferably use a piece of squared paper, on which this is already printed. Start with some value of x, say $x = 1$; the corresponding value of y is $y = 2$. This is shown on the graph by making a dot one unit to the right of the up-and-down line, and two units above the across line. Next let $x = 2$; then $y = 3$, which fixes the next dot. Then $x = 3$ gives $y = 4$, which fixes the next one and so on. It is easily seen that all these dots lie in a straight line, and this straight line is the graph of the function $y = x + 1$.

The reader should draw in a similar way the graphs of other simple functions, such as $y = 2 - x$, or $y = x^2 - 1$.

What we are doing here is simply the converse of the process used in Chapter V. There we wished to bring geometrical straight lines and curves within the scope of calculation, and we did this by labelling points by pairs of numbers, and finding relations between these numbers, which we called the equations of the straight lines or curves. Here we start with the equations, and get a pictorial representation of them by representing the number-pairs as dots on paper, and the equations as curves drawn on the paper.

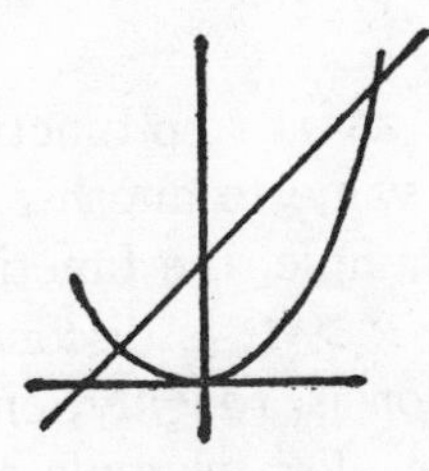

It is possible to use graphs to get rough solutions of algebraic equations. Suppose, for example, that in the same diagram as has been used to draw the graph of $y = x + 1$, we also draw the graph of $y = x^2$. To do this at all accurately points much closer together than $x = 1, 2, \cdots$ must be used. Some idea of the shape of the graph may be obtained by plotting the points corresponding to $x = .1, .2, \cdots$. The curve corresponding to this equation is a parabola whose appearance is roughly that of the curved line in the above diagram. It will be seen that the straight line intersects the parabola in two points. At each of these points y is equal to $x + 1$, and it is also equal to x^2. Consequently $x^2 = x + 1$ for the two values of x corresponding to these points. If the graph has been drawn very accurately, the approximate values of x can be found by measurement from the graph. The values found by solving the equations by theoretical methods are $x = \frac{1}{2} + \frac{1}{2}\sqrt{5}$ and $x = \frac{1}{2} - \frac{1}{2}\sqrt{5}$. Since $\sqrt{5} = 2.236$ approximately, these are approximately equal to 1.618 and $-.618$. It would however need a very large scale and accurate drawing to get such a good approximation from the graph. In C. Smith's *Algebra*, a well-known mathematical text-book of the last age, I find the remark "Graphical methods are, however, after all only methods of getting rough approximations by those who know no Algebra." This seems rather severe, as there are many equations which it is beyond the power of algebra to solve exactly.

Continuous functions.

A function such as a step-function which jumps suddenly from one value to another is said to be discontinuous. For example, the function equal to 1 for $x \leqslant 1$, and to 2 for $x > 1$, is discontinuous at $x = 1$. A continuous function is, roughly, one which does not do this sort of thing. For example the function equal to x for $x \leqslant 1$, and to x^2 for $x > 1$, is continuous at $x = 1$, in spite of the change of formula, because the two pieces join up at this point.

One thinks of the graph of a continuous function as a line which can be drawn on paper without lifting up the pencil. This gives some idea of the situation, if we remember that we must not join up the steps at discontinuities by lines parallel to the y-axis. Such a line would not represent anything in the nature of the function, which can only have one value for each value of x.

The official definition of continuity is that a function

is continuous if its value at any point is the limit of its values as the independent variable approaches this point. It is not very obvious how to apply this, but it does give the right sort of effect; for example, it ensures that a continuous function cannot pass from one value to another without passing through all intervening values. It might be thought that this property could be taken as the definition, but it turns out that it would not be at all satisfactory.

Periodic functions.

In the chapter on trigonometry we considered the values of the function $y = \sin x$, where x represents an angle not greater than a right angle; measured in radians, x lies between 0 and $\frac{1}{2}\pi$. As x increases steadily from 0 to $\frac{1}{2}\pi$, $\sin x$ increases steadily from 0 to 1. This

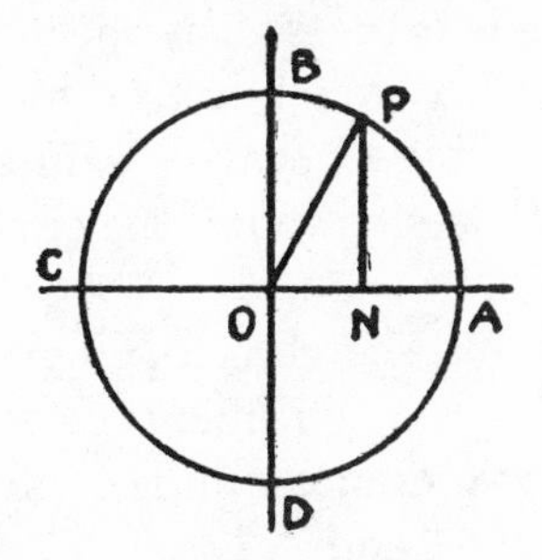

is easily seen from a figure, in which a point P travels round a circle of radius 1, with centre O, say. If P starts from the position A, POA is the angle of circular measure x, and PN is perpendicular to OA, then $\sin x$ is equal to PN. When P is at B, $x = \frac{1}{2}\pi$ and $\sin x = BO = 1$.

Now suppose that P travels on round the circle beyond B. The value of PN, i.e., of $\sin x$, decreases again, and when P is at C it vanishes. We express this by means of the formula $\sin \pi = 0$.

Now let P go still farther round, beyond C. It is now below the line AC, and PN is in the opposite direction to what it was before. It is natural to attach the negative sign to it in this case. The result is that, if x lies between π and 2π, $\sin x$ is negative. When P is at D, $x = \frac{3}{2}\pi$, and consequently $\sin \frac{3\pi}{2} = -1$. When P gets round to A again, x is 2π, and so $\sin 2\pi = 0$.

Beyond this, the situation simply repeats itself. As x goes from 2π up to 4π, P passes round the circle once more, and $\sin x$ goes through the same cycle of values as it did when x went from 0 to 2π. Similarly when x goes from 4π to 6π, and in fact the same thing goes on repeating itself indefinitely.

A function with this property is called a periodic function. The interval through which x has to pass in order that the function y shall pass through its whole cycle of values is called the period. Thus $\sin x$ has the period 2π. This is expressed by means of the formula $\sin (x + 2\pi) = \sin x$, which is true for all values of x. We can of course give x negative values, corresponding in the figure to P passing round the circle from A in the opposite direction.

The reader should sketch a rough graph of the function $y = \sin x$. It is a continuous curve, starting from the origin of co-ordinates and oscillating endlessly between the values 1 and -1.

All the functions of trigonometry also have similar periodic properties. The graph of the function $y = \cos x$, for example, is quite similar to that of $\sin x$. In fact it can be looked on as the same curve, moved along through a distance $\frac{1}{2}\pi$ in the direction of the axis of x. This is expressed by means of the formula $\sin (x + \frac{1}{2}\pi) = \cos x$.

The function $\tan x$ is also periodic, its period being equal to π; but its graph has quite a different appearance, since it goes off to infinity as x approaches the value $\frac{1}{2}\pi$, and then reappears from infinity in the opposite direction.

The property of periodicity is of importance in many of the applications of mathematics. Many of the functions of physics are at any rate roughly periodic; everything which has a daily variation, for example, is periodic, with a period of 24 hours. In the representation of such variables, periodic functions are obviously required.

Tables.

Another way in which a function can be represented approximately is by means of a table. A table is a list of the values which the function takes for certain values of the independent variable. Suppose for example that we wish to tabulate a function $y = f(x)$ for values of x between 0 and 1. We have to begin by selecting values of x which are sufficiently close together to give a good idea of how the function behaves. We might for example take $x = .01, .02, \cdots .99, 1$. The table would then contain 100 entries. Whether this would be sufficient would depend entirely on the purpose for which the table was required, and on the nature of the function to be tabulated. A rapidly varying function would require more entries in the table than a slowly varying one, to give the same degree of accuracy.

Having selected the values of x, we should have to calculate the corresponding values of y, to a certain number of decimal places; and these values are the entries which appear in the table. Here again we should have to decide from practical considerations how many decimal places to use. The more we use, the more accurate the table will be, but the more space it will occupy, and the greater will be the labour of calculating it.

An example of a table has been given on page 136. This is an extract from a table of the function $y = \sin x$. The whole table goes up to $x = 90°$, and each degree is divided into six equal parts, so that there are 540 entries in the table.

At each value of x, y is calculated to five decimal places. In this case, y is a steadily increasing function of x; that is, as we pass along the table from any value of x to the next greater value, the corresponding value of y also increases. But in other cases y might decrease, or it might sometimes increase and sometimes decrease.

In a sense, a table does the same thing as a graph; it gives an approximate representation of a function by giving approximations to its values at certain points. But a table can be made very much more accurate than any graph.

I happen to have here Milne-Thomson's *Standard Table of Square Roots*. As an example, let us use it to solve approximately the same equation as before, $x^2 = x + 1$, without working out its theoretical solution. This equation is equivalent to $x = \sqrt{(x + 1)}$, or, if we put $x + 1 = t$, to $t - 1 = \sqrt{t}$; that is, we have to find a number whose square root is one less than the number itself. Looking down the table, I find the following entries (I have adjusted the decimal point as required here)

t	$\sqrt{t}$
2.617	1.6177144
2.618	1.6180235
2.619	1.6183325

The number we want is clearly $t = 2.618$, corresponding to $x = 1.618$. It is here that the graphs of $t - 1$ and $\sqrt{t}$ would cross, if we drew them as accurately as the tables would allow. The table shows, of course, that this is not an exact solution, but it is the nearest that can be got from the table. The last three or four decimal places to which $\sqrt{t}$ is calculated are of no use for this purpose.

The uses of tables in trigonometry have already been described.

Functions of several variables.

In each of the above examples of functions there was just one independent variable, x, t or whatever it might be, on which the function depended. But a function may depend on two, three, or indeed any number of independent variables. There are many examples in physics of such functions. The force exerted by two gravitating bodies on each other is a function of the masses of the bodies as well as of the distance between them. The formula which expresses this dependence is $mm'g/r^2$, where m and m' are the masses, r is the distance and g is a constant. To take an example from trigonometry, the cosine of any angle of a triangle is a function of the three sides of the triangle. The formula which expresses the dependence of $\cos A$ on a, b, and c was given in Chapter XI.

The general notation for a function of two variables x and y is $f(x, y)$, and similar expressions are used in the case of more variables.

We cannot draw on paper the graph of a function of two variables. This would require a surface in three dimensions, not just a curve. If the function is $z = f(x, y)$, and x, y and z are thought of as the coordinates of a point in three-dimensional space, then the equation will represent a surface. Models of such surfaces are sometimes made with plaster or with string, but so far as I know these are not of any practical use.

Chapter XIII

THE DIFFERENTIAL CALCULUS

The speed of a moving body may be measured by the number of miles per hour, or of feet per second, which it goes. In any units the speed is the number of units of distance traversed during the motion, divided by the number of units of time taken. Now this is all right as long as the body is going steadily along, so that its motion does not vary from instant to instant. But if the motion is itself changing, the total distance divided by the total time will not give any idea of what is happening at any particular instant. It will only give a sort of average speed over the whole motion.

To express this in formulae, suppose that, at a time t, a moving body has reached a distance x from some starting-point. Suppose that, at a later time t', it has reached a distance x'. Then "distance divided by time" gives the formula $\frac{x' - x}{t' - t}$. If the motion is steady, this can be called the speed or velocity of the body. If not, it is only an average velocity, and it gives no idea of what the actual velocity will be at any particular instant (say at the time t), or indeed whether such a velocity at an instant can be defined at all.

If a fast car went by you, and you could take the time at which this occurred, and the time at which it reached a point half a mile up the road, then, on dividing distance by time, you would get a number, which you might think had some relation to the speed with which it passed you. It might be better to time it over a quarter of a mile, provided that the timing was accurate enough. To time it over a few yards would probably be useless, as the slight vagueness inherent in the timing would make the result very inaccurate. To time it over an inch or so would be obviously futile. The same sort of thing is true of any observation of physical quantities. But, for a mathematical function defined by a formula, this difficulty does not occur.

Suppose we think of a body falling under gravity; and then not really of the falling body, but of the mathematical model by which we represent it to ourselves. In the model, let x be the distance through which it falls from rest in time t. Then the formula connecting x with t is $x = \frac{1}{2}gt^2$, where g is a constant (the constant of gravitation). We then want to find the velocity of the "body" at a given instant t.

Let us time it over an interval beginning at this instant. Let us call the time at the end of this interval $t + h$, so that h units of time have elapsed during the interval. At the beginning of the interval the body has fallen a distance $\frac{1}{2}gt^2$, and at the end a distance $\frac{1}{2}g(t + h)^2$. During the interval it therefore goes a distance

$$\tfrac{1}{2}g(t + h)^2 - \tfrac{1}{2}gt^2,$$

but

$$\begin{aligned}\tfrac{1}{2}g(t + h)^2 &= \tfrac{1}{2}g(t^2 + 2th + h^2)\\ &= \tfrac{1}{2}gt^2 + gth + \tfrac{1}{2}gh^2.\end{aligned}$$

The distance traversed is therefore $gth + \frac{1}{2}gh^2$. The "distance divided by time" rule thus gives $gt + \frac{1}{2}gh$. This may be called the average velocity over the whole interval.

The expression $gt + \frac{1}{2}gh$ which we have thus found consists of two parts. The term gt depends only on the instant t at which we start the measurement, and on the constant g, but not on the length of time h over which the motion takes place. The second term, $\frac{1}{2}gh$, does depend on h; but it is clear that the smaller h is, the smaller this term will be. Suppose the experiment performed successively with smaller and smaller values of h, say for example that each is half of the one before. Then this term also decreases indefinitely, or, as we say, tends to zero. The whole expression then approximates as closely as we like to the value gt.

It is clear that this number gt has some special significance in the problem. It is the limit of the average velocity over the interval $(t, t + h)$, when h is made indefinitely small. It seems fair then to call it *the velocity at the time t.* This is the definition of what we mean by "velocity at an instant" in this particular case. Naturally "velocity at an instant" can be defined similarly in many other cases. It does not follow that it exists in every conceivable case, but it does exist if the motion is given by any of the simplest functions of mathematics.

The differential calculus.

The reader who has followed the above argument about the velocity at an instant of the idealized falling body has performed the first step in the differential calculus. This celebrated branch of mathematics was invented independently by Newton and Leibnitz in the 17th century. Their claims to priority were the subject of a famous controversy at the time, but we need not attempt to judge the matter now. The differential calculus is such a necessary part of mathematics that we may suppose that sooner or later it would have been discovered by someone. Most other branches of mathematics now depend on it in some way or other.

The differential calculus is primarily concerned with the measurement of the rates of change of things which do change. In mathematics the things which change are functions of a variable. In the above example, we determined the rate of change of the function $x = \frac{1}{2}gt^2$.

We might ask the same question about a still simpler function, the function $x = gt$. The value of this at a later time $t + h$ is $g(t + h)$. The change during this interval is therefore gh, and on dividing by h we obtain simply g. This measures the rate of change of the function gt. In this case, as it happens, it does not matter what value of h we take, and the result is the same whether h is large or small.

This operation of finding the rate of change of a given function is known as differentiation; we say in cases such as those given above that we differentiate x with respect to t. There are various notations by which this is expressed. Suppose that the function concerned is denoted by $x = f(t)$. Then the rate of change at the instant t is denoted by $f'(t)$. In general this also will depend on t, so that $f'(\ )$ is just another functional symbol, like $f(\ )$. Thus if $f(t) = \frac{1}{2}gt^2$, then $f'(t) = gt$, and if $f(t) = gt$, then $f'(t) = g$. In the latter case, the second function happens to be a constant.

Another rather peculiar notation consists of writing $\frac{dx}{dt}$ instead of $f'(t)$. This is to be read "*dx* by *dt*." It is not a fraction, and the reader should not attempt to take it to pieces and attach a meaning to each separate piece. It just means the same thing as $f'(t)$. It has the advantage that it contains in a kind of petrified form both the t and the x, and so reminds us which is the independent variable, and which is the dependent variable. The origin of such expressions is as follows. Instead of writing t and $t + h$ as values of the independent variable at the ends of the interval considered, people often used to write t and $t + \delta t$; here δ is not a number multiplying t, but δt as a whole means the same thing as h. It is the change in t. The corresponding change in x is then denoted by δx (in the first example which we considered, δx was equal to $gt + \frac{1}{2}gh$). The average rate of change over the interval is then $\frac{\delta x}{\delta t}$, this being a fraction. The instantaneous rate of change is got from this by making δt tend to zero. We write it as $\frac{dx}{dt}$, but this is not a fraction, and the dx and dt do not have separate existences.

Such expressions have nothing to do with the fraction $\frac{0}{0}$, which we agreed in Chapter IV to exclude from our scheme of things. The idea that it has somehow crept back here would be quite wrong; $\frac{dx}{dt}$ is not "dx divided by dt," but merely a conventional notation for the limit of "δx divided by δt."

The $f'(t)$ or $\frac{dx}{dt}$ is called the differential coefficient of x with respect to t, or the derived function, or just the derivative.

As another example of a differential coefficient, consider the function $y = \sin x°$ (measured in degrees), a table of which is given on page 136. Let us see what this table tells us about the rate of change of the function at the value $x = 30°$. The corresponding value of y is .5. Let us start by taking another value of x not too near to the first, say for example $x = 31°$. Comparing this with the first value, the change in x, or δx as we may call it, is 1 degree. The corresponding change in y, which we call δy, is the difference between the entry in the table for $x = 31°$, and that for $x = 30°$, and so is .01504. Hence

$$\frac{\delta y}{\delta x} = \frac{.01504}{1} = .01504.$$

If we take instead the next smaller value of x, $x = 30° \ 50'$, i.e., $x = 30\frac{5}{6}°$, then $\delta x = \frac{5}{6}$, and, from the table, the corresponding value of δy is .01245. Hence

$$\frac{\delta y}{\delta x} = \frac{.01254}{\frac{5}{6}} = .01505.$$

Similarly $x = 30° \ 40'$ gives

$$\frac{\delta y}{\delta x} = \frac{.01004}{\frac{4}{6}} = .01506,$$

$x = 30° \ 30'$ gives

$$\frac{\delta y}{\delta x} = \frac{.00754}{\frac{3}{6}} = .01508,$$

$x = 30° \ 20'$ gives

$$\frac{\delta y}{\delta x} = \frac{.00503}{\frac{2}{6}} = .01509,$$

and $x = 30° \ 10'$ gives

$$\frac{\delta y}{\delta x} = \frac{.00252}{\frac{1}{6}} = .01512.$$

Running our eye down this set of values of $\delta y/\delta x$, we see that they increase steadily as δx decreases. This is some indication that they may be examples of a sequence which tends to a limit as $\delta x \rightarrow 0$, though the table does not allow us to follow the sequence any farther. The value of the limit, i.e., of $\frac{dy}{dx}$ may be expected to be somewhere in the neighbourhood of .0151.

This is confirmed by theory; for it is known that, if $y = \sin x$, when x is measured in radians, then $\frac{dy}{dx} =$

$\cos x$. At the value of x considered, $\sin x = \frac{1}{2}$, and so

$$\cos x = \sqrt{1 - \sin^2 x}$$

$$= \sqrt{1 - \frac{1}{4}} = \sqrt{\frac{3}{4}} = \frac{\sqrt{3}}{2} = \frac{1.732}{2} = .866.$$

Now π radians correspond to 180°; consequently if the x-unit is a degree instead of a radian, we shall have to divide by 180 and multiply by π. The value of $\frac{dy}{dx}$ in these units will therefore be approximately $.866 \times 3.14 \div 180$, and this is just about .0151.

It will be noticed that the values of $\delta y/\delta x$ increase fairly steadily, but apparently with rather a jump at the last stage. This is due to the fact that here the denominator is getting rather small. Here we are dividing by $\frac{1}{6}$, i.e., multiplying by 6. Now the table is only "correct to five decimal places," so that there may be an error of anything up to .000005. On multiplying by 6, such an error would turn into .00003, and so the last decimal place in $\delta y/\delta x$ may be out to this extent. This difficulty is of course inherent in all numerical approximations to a differential coefficient. The smaller δx is, the more accurately you have to observe y in order to get a good approximation to $\frac{dy}{dx}$.

In the differential calculus we often use the language of variation in time, rate of change, and so on, but the subject has no necessary connection with time. The independent variable is often denoted by x, and the dependent variable by y, and (x, y) is thought of as a point in Cartesian geometry. The differential coefficient must then have a geometrical meaning. We shall next see what this is.

The function $y = f(x)$ corresponds to a curve in the

plane, and (x, y) is a point, P say, on the curve. Now change x into $x + \delta x$, and let the corresponding value of y be $y + \delta y$. The point $(x + \delta x, y + \delta y)$ is another point, Q say, a little farther along the curve. We can

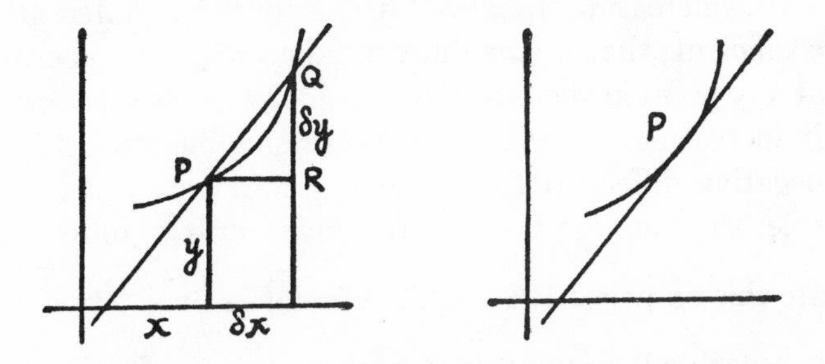

make a little triangle PQR by drawing PR parallel to the x-axis, and QR parallel to the y-axis, meeting at R. The angle at R is a right angle, and the length of PR is δx, and that of QR is δy. Now use the notation of trigonometry: we have

$$\frac{\delta y}{\delta x} = \tan \mathrm{P},$$

i.e., the tangent of the angle at P. This angle is equal to the angle which the chord PQ makes with the axis of x.

As we make δx, and so also δy, smaller and smaller, the point Q slides along the curve towards P. To say that the differential coefficient $\frac{dy}{dx}$ exists at P is equivalent to saying that, as Q approaches P, the chord PQ approaches a limiting position, which will be simply that of the tangent (in the geometrical sense) to the curve at P. The geometrical meaning of $\frac{dy}{dx}$ is then, that it is the tangent (in the trigonometrical sense) of the angle which the tangent (in the geometrical sense) to the curve at P makes with the axis of x.

We can think of this as a measure of the steepness or

gradient of the curve at P. The gradient may be defined as "amount of rise per unit of horizontal distance," and this is just what $\frac{dy}{dx}$ is.

An increasing function has a positive differential coefficient; that is, the differential coefficient is positive at any point in the neighbourhood of which the function is increasing. Similarly, a decreasing function has a negative differential coefficient. As an example, consider the function $y = x^2$. We have learnt to differentiate this on pages 153–54. The result is $\frac{dy}{dx} = 2x$, which is negative if x is negative, and positive if x is positive; this obviously corresponds to the fact that x^2 decreases as x increases through negative values (e.g., it decreases from 100 to 0 as x increases from -10 to 0), and then increases again as x increases through positive values.

If we mount a smooth rounded hill to the top, pausing there to take breath, we seem to be for the moment on level ground. Here the gradient ceases to be positive, and as we go on it becomes negative again. In mathematical language, $\frac{dy}{dx}$ changes from positive to negative at a maximum. If there is a definite gradient just at the top, it must be zero, so that $\frac{dy}{dx} = 0$. There is a similar state of affairs at the bottom of a valley, where the gradient changes from negative to positive, and vanishes just at the bottom. A general test for the maximum or minimum of functions is therefore that the differential coefficient must vanish at such points. An example of a minimum is given by the above function $y = x^2$. This clearly has a minimum at $x = 0$, and in fact $\frac{dy}{dx} = 2x = 0$ there. As an example of a maximum, consider the function $y = x - x^2$. Changing x into

$x + h$, y is changed into $x + h - (x + h)^2 = x + h - x^2 - 2xh - h^2$. Subtracting the original value of y, we obtain $h - 2xh - h^2$, and, dividing by h, the result is $1 - 2x - h$. This consists of the usual two parts, the $1 - 2x$ independent of h, and the remainder $-h$ which tends to zero as h tends to zero. Consequently the differential coefficient is $1 - 2x$. This vanishes when $x = \frac{1}{2}$. The corresponding value of y is $\frac{1}{2} - \frac{1}{4} = \frac{1}{4}$, and this is a maximum. It can be seen in various ways that it is the greatest value which y can have, for any value of x.

Higher differential coefficients.

If a function $y = f(x)$ has a differential coefficient $f'(x)$, this itself will be a function of x. If it is the right sort of function, it will itself have a differential coefficient; this is denoted by $f''(x)$; and so on, it being possible in some cases to go on forming differential coefficients indefinitely. In the other notation, the first differential coefficient is denoted by $\frac{dy}{dx}$, the second by $\frac{d^2y}{dx^2}$, the third by $\frac{d^3y}{dx^3}$, and so forth; but the reader is recommended not to try to dissect these formulae into their component parts. As an example, if $y = x^2$, then $\frac{dy}{dx} = 2x$, $\frac{d^2y}{dx^2} = 2$, $\frac{d^3y}{dx^3} = 0$, and naturally all the following differential coefficients are 0.

The second differential coefficient has a fairly simple geometrical meaning. It is the gradient of the gradient. If it is positive, then the gradient is increasing, a situation such as we experience just after passing a valley bottom to mount the opposite hill. Similar meanings can be given in other cases. The meanings of the higher differential coefficients are naturally more complicated.

If we have a table of values of a function, we can tell from it something about the differential coefficients of the function. Consider again the table of the function $y = \sin x°$ on page 136. This is an increasing function, so that the gradient or first differential coefficient $\frac{dy}{dx}$, will be positive in this interval. Now consider the successive differences between the entries in the table. They are .00252, .00251, .00251, .00250, .00250, .00250, .00249, $\cdots$. There is very little difference between them, but they definitely get less, so that the gradient is decreasing, and the second differential coefficient is negative. The number of decimal places used in the table is not enough to tell us anything definite about the higher differential coefficients.

A constant may be regarded as a simple particular case of a function. If $y = 1$, say, whatever x is, it may be convenient to think of y as a function of x, but it is a function with only one value. In the Cartesian diagram, such a function is represented by a straight line parallel to the x-axis. The gradient of such a line is zero. It is easily seen directly that if $y = c$, a constant, then $\frac{dy}{dx} = 0$; for the change δy in y corresponding to any change δx in x is 0, so that $\delta y/\delta x = 0$, and so its limit $\frac{dy}{dx}$ is also 0.

Now a very important theorem of the differential calculus asserts that the converse of this is also true; if the differential coefficient of a function is always 0, then the function must be a constant. It may be thought that this is rather obvious. A road whose gradient is always nil can never go either up or down, and so must remain completely flat. This is common sense, but it is not mathematics; and in mathematics the guidance of common sense is not always quite reliable. A logical deduction from the data is something different from such commonsense arguments. The proof is not difficult, but it would be out of place here. We can only assure the reader that in this case mathematics and common sense reach the same conclusion.

This theorem has some important consequences. Many mathematical problems take a form in which we know something about the differential coefficients of a function, and want to find out what the function itself is. An elaborate technique has been worked out for solving problems of this kind. Let us just see what we can say about a simple example. Suppose that there is a function y of x, and all that we know about it is that $\frac{dy}{dx} = 1 - 2x$ for all values of x. What is y? As it happens we have just been dealing with a function with this property, namely $y = x - x^2$. Are there any other solutions of the problem? Yes; if $y = x - x^2 + c$, where c is any constant, then y has the required property; and there are no other solutions of any kind. For if there are two solutions, y and z, then $\frac{d}{dx}(z - y) = \frac{dz}{dx} - \frac{dy}{dx} = 0$, since $\frac{dz}{dx}$ and $\frac{dy}{dx}$ are both equal to $1 - 2x$. Hence $z - y$ is a constant, say c. Taking y to be the function $x - x^2$ which we started with, it follows that $z = x - x^2 + c$. Thus all possible solutions are comprised in this formula.

The solution of the problem contains a constant c which is not determined by anything which we have yet supposed to be given. It is what we call an arbitrary constant; that is, we still have it at our disposal, and we can choose it so that the solution fits in with one more condition, imposed, for example, by the necessities of a physical problem. Such a condition usually takes the form of saying that, when x has a particular value (say for example $x = 0$), then y also is to have a prescribed value (say for example $y = 1$). In the above case, this means that the equation $y = x - x^2 + c$ is to be satisfied if $x = 0$ and $y = 1$; and this plainly requires that $c = 1$. The solution is then completely determined —it is $y = x - x^2 + 1$.

From the geometrical point of view, the situation is now as follows; the equation $\frac{dy}{dx} = 1 - 2x$ does not determine completely the curve in the Cartesian plane which has this property; but if we add that the curve must pass through the point $(0, 1)$, then it is completely determined.

Differential equations.

The above problem is one of the simplest examples which can be given of what is called a differential equation. A differential equation is an equation connecting an independent variable x with a dependent variable y, but containing also the differential coefficient of y with respect to x, and possibly also the second differential coefficient, and even differential coefficients of still higher order. The above equation, $\frac{dy}{dx} = 1 - 2x$, is called a differential equation of the first order, because it contains nothing worse than $\frac{dy}{dx}$. As an example of a differential equation of the second order, we may give the equation $\frac{d^2y}{dx^2} + y = 0$. Constructing a relation between x and y from which this could be obtained by differentiation is known as solving the differential equation. It would be impossible to explain here how to solve the last equation which we have written down; it must be sufficient to say that it has the solutions $y = \sin x$ and $y = \cos x$, where x is expressed in radians.

In the case of a second-order differential equation, it is possible to find a solution, which not only passes through a given point in the (x, y) plane, but passes through it in a given direction. For example, we can find a solution of the above equation which passes through the point (0, 1), and has there the gradient 1. It is actually $y = \sin x + \cos x$, but I must refer to books on differential equations for the method of obtaining this result.

Such problems are of frequent occurrence in mechanics, that is in the study of the mathematical models by which we represent to ourselves the motions of material bodies. The science of mechanics was developed by Galileo, Newton and others in the 16th and 17th centuries. In it, a body is supposed to be subject to certain forces, which affect its motion according to certain laws. Expressed in the language of mathematics, these laws usually take the form of differential laws; that is, they connect the position, velocity, acceleration, etc., of the body at a particular instant. They do not primarily tell us the whole motion, but merely the differential laws governing it. It is the motion as a whole which has to be deduced from the differential law. In other words, it is a question of solving differential equations, in which the time is the independent variable, and there are one or more dependent variables which give the position of the body.

Such, for example, is the problem of the motion of a planet round the sun. According to Newton's theory of gravitation, a planet is subject to a force of attraction towards the sun, which is inversely proportional to the square of its distance from the sun. This means that the x, y, or whatever they may be which determine the position of the planet satisfy certain differential equations, with the time t as independent variable. If the planet is thought of as a particle which is shot off from a given point in a given direction, its path will be determined ever afterwards by the theory of differential equations. The result is actually the famous law of

Kepler, that the planet will go on endlessly round an ellipse, of which the sun is a focus.

Similar laws determine what takes place when an apple falls from a tree, only in this case the motion is soon brought to an abrupt end by a collision between the apple and the earth.

These theories gave rise at one time to the idea that, if we could determine the position and motion of every particle in the universe at any one instant, we could predict the motion of the whole system afterwards to all eternity. Such ideas ignore the difference between actual physical systems and the mathematical models which we make of them. There is no reason to identify the two things down to the last detail, which, in the physical case, must presumably always remain unknown. Consequently the mathematical model should never be taken too seriously, in its physical applications.

Successive approximations.

A very important idea in mathematics is that of the solution of problems by means of successive approximations. The idea is that we first find an approximate solution of a problem; then, on the basis of this, we look for a still nearer solution, or, as we say, a second approximation; and so on, each time getting a little nearer to the exact solution. This process takes many different forms. An example occurs in the following problem.

Suppose that we are given a function, $y = f(x)$ say, which has differential coefficients of as many orders as may be required. Suppose that we know its value for a particular value of x, and also the value of its differential coefficients $f'(x), f''(x), \cdots$ for this value of x. Can we find the value of the function when x is replaced by a slightly greater value $x + h$? What "slightly greater" means depends on the circumstances of the problem, but we shall see how it works out without being too precise about this.

If $f(x)$ is a continuous function, naturally $f(x+h)$ is nearly equal to $f(x)$ if h is small. This means that $f(x)$ is a first approximation to $f(x+h)$. We may write this as

$$f(x+h) = f(x) + \cdots$$

where the $+ \cdots$ means that the equation is not exact, and that to make it exact something more, not yet determined, would have to be put in.

To obtain a second approximation, we recall the definition of a differential coefficient; it is that $f'(x)$ is the limit of $\dfrac{f(x+h) - f(x)}{h}$ as h tends to 0. Consequently $\dfrac{f(x+h) - f(x)}{h}$ will be approximately equal to $f'(x)$ when h is small, and so $f(x+h) - f(x)$ will be approximately equal to $hf'(x)$. This is just what we want; it shows that the next approximation is

$$f(x+h) = f(x) + hf'(x) + \cdots$$

where the $+ \cdots$ means again that there is still something lacking to exact equality.

A third approximation is not quite so easy to find, but the following considerations may suggest it. In the above formula we have gone in one jump from x to $x + h$, with some sacrifice of accuracy. It would be more accurate to follow the curve $y = f(x)$ round by means of the intervening points $x + \dfrac{h}{n}$, $x + \dfrac{2h}{n}$, $\cdots$ $x + \dfrac{(n-1)h}{n}$, where n is a large positive integer, and to calculate approximately the successive changes in y. We could do this by means of the "second approximation" already found. This gives in the first place

$$f\left(x + \frac{h}{n}\right) - f(x) = \frac{h}{n} f'(x) + \cdots,$$

then

$$f\left(x+\frac{2h}{n}\right)-f\left(x+\frac{h}{n}\right)=\frac{h}{n}f'\left(x+\frac{h}{n}\right)+\cdots,$$

$$f\left(x+\frac{3h}{n}\right)-f\left(x+\frac{2h}{n}\right)=\frac{h}{n}f'\left(x+\frac{2h}{n}\right)+\cdots$$

until finally

$$(x+h)-f\left(x+\frac{(n-1)h}{n}\right)$$
$$=\frac{h}{n}f'\left(x+\frac{(n-1)h}{n}\right)+\cdots.$$

We can also apply the same rule to the function $f'(x)$ instead of $f(x)$. This gives

$$f'\left(x+\frac{h}{n}\right)=f'(x)+\frac{h}{n}f''(x)+\cdots,$$

$$f'\left(x+\frac{2h}{n}\right)=f'(x)+\frac{2h}{n}f''(x)+\cdots$$

and so on, until finally

$$f'\left(x+\frac{n-1}{n}h\right)=f'(x)+\frac{n-1}{n}hf''(x)+\cdots$$

Using these expressions, the above formulae become

$$f\left(x+\frac{h}{n}\right)-f(x)=\frac{h}{n}f'(x)+\cdots,$$

$$f\left(x+\frac{2h}{n}\right)-f\left(x+\frac{h}{n}\right)=\frac{h}{n}f'(x)+\frac{h^2}{n^2}f''(x)+\cdots$$

$$f\left(x+\frac{3h}{n}\right)-f\left(x+\frac{2h}{n}\right)=\frac{h}{n}f'(x)+\frac{2h^2}{n^2}f''(x)+\cdots,$$

and so on, until finally

$$f(x+h) - f\left(x + \frac{n-1}{n}h\right)$$
$$= \frac{h}{n}f'(x) + \frac{n-1}{n^2}h^2f''(x) + \cdots.$$

Now add all these expressions together. On the left-hand side, all the terms cancel in pairs except the $f(x)$ in the first formula and the $f(x+h)$ in the last, so that the sum is $f(x+h) - f(x)$. On the right-hand side, $\frac{h}{n}f'(x)$ occurs n times, giving the sum $hf'(x)$. Altogether we obtain

$$f(x+h) - f(x) = hf'(x)$$
$$+ \left\{1+2+\cdots+(n-1)\right\} \frac{h^2}{n^2} f''(x) \cdots$$
$$= hf'(x) + \tfrac{1}{2}n(n-1)\frac{h^2}{n^2} f''(x) + \cdots$$

on using the formula for the sum of an arithmetical progression.

The expression which multiplies $f''(x)$ in this formula is equal to $\frac{1}{2}h^2 - \frac{1}{2}\frac{h^2}{n}$, and the last term will be negligible if n is large. Hence finally we obtain the formula

$$f(x+h) = f(x) + hf'(x) + \tfrac{1}{2}h^2f''(x) + \cdots$$

as our third approximation.

The above argument is open to a good deal of criticism. In every formula there is a $+ \cdots$, which means that there is something left over; and we have treated these somethings as if they were really not there at all. Actually a certain vagueness is inevitable in the argument, because we began by being rather vague about the properties of the function $f(x)$ with which we were dealing. It requires a good deal of mathematical training to put this sort of thing on a firm basis. But it can be done, and the formulae obtained really do give good approximations to reasonable functions for suitable values of x and h.

In some of the previous sections we assumed that a function $\sin x$ had been tabulated, and we used the table to find out something about the differential coefficients of the function. Actually of course we should go the other way about, and deduce the values in the table from a theoretical knowledge of the function. It may be interesting to see how this works out in a particular case. Suppose that we wish to find approximately the value of the sine of $30°\ 10'$, i.e., of $30\frac{1}{6}°$. To express this in radians, we have to multiply by $\dfrac{\pi}{180}$, since there are π radians in $180°$. It is therefore a question of evaluating $\sin\left(\dfrac{\pi}{6} + \dfrac{\pi}{1080}\right)$ approximately. Now this is an expression of the form considered in the previous section. The function $f(\ )$ is $\sin(\ )$, the number x is $\frac{1}{6}\pi$, and h, the small addition to x, is $\dfrac{\pi}{1080}$. As we have already mentioned, the differential coefficient of $\sin x$ is $\cos x$. Hence if we use the formula which we called the second approximation, we obtain

$$\sin\left(\frac{\pi}{6} + \frac{\pi}{1080}\right) = \sin\frac{\pi}{6} + \frac{\pi}{1080}\cos\frac{\pi}{6} + \cdots.$$

The value of $\sin\dfrac{\pi}{6}$ is $\frac{1}{2}$, or .5; that of $\cos\dfrac{\pi}{6}$ is very nearly .866; and π is nearly 3.1416, which gives .00291 as the approximate value of $\pi/1080$. On doing the arithmetic it is found that the right-hand side of the above expression is very nearly .50252. This is just the value which is given in the table on page 136.

We could of course obtain the value to more decimal places by using the third approximation obtained in the previous section. In doing so we should also have to use more decimal places in the approximate values of π and $\cos \frac{\pi}{6}$.

It may be asked what the above argument really proves about the value of $\sin \left(\frac{\pi}{6} + \frac{\pi}{1080}\right)$. The answer is that strictly it proves nothing. It is all a question of the $+ \cdots$. We have assumed that this represents a number which is negligibly small, but it still remains to prove that it is so. This is a job for mathematicians, and it would take too long to explain here how it is done. However, it can be done, and the result obtained above can be justified without much difficulty.

Taylor's theorem.

It would be possible to continue the process outlined in the above section so as to obtain still further approximations, but the details would be rather complicated. A more promising line of investigation is to examine the general form of the above formulae, and to see what that suggests. Now in the terms $hf'(x)$, $\frac{1}{2}h^2f''(x)$ which we have added on to the right-hand side, the power of h goes up by one each time. This suggests strongly that the next term will be of the form Ch^3, where C is something which does not depend on h. The formula suggested is thus

$$f(x + h) = f(x) + hf'(x) + \tfrac{1}{2}h^2f''(x) + Ch^3 + \cdots.$$

It is then a question of finding what C must be. Briefly the method is to differentiate repeatedly with respect to h, keeping x fixed. The successive results of this are

$$f'(x + h) = f'(x) + hf''(x) + 3Ch^2 + \cdots,$$

$$f''(x + h) = f''(x) + 6Ch + \cdots$$

and

$$f'''(x + h) = 6C + \cdots$$

the $+ \cdots$ at each stage indicating an expression containing higher powers of h. On putting $h = 0$, all the terms on the right-hand side vanish except the first one in each case. The first two equations reduce to identities, which merely confirm what we have already learnt. The last one reduces to $f'''(x) = 6C$. Hence $C = \frac{1}{6}f'''(x)$, and the fourth approximation is

$$f(x + h) = f(x) + hf'(x) + \tfrac{1}{2}h^2f''(x) + \tfrac{1}{6}h^3f'''(x) + \cdots.$$

This suggests that we have got on to something which goes on indefinitely according to a fairly simple rule. The rule becomes clear if we notice that the factors 2, 6, in the denominator are equal to the factorials 2!, 3!. It is therefore suggested that there is a formula which is not just an approximation consisting of a finite number of terms, but an infinite series whose sum is actually equal to the function with which we start. The formula will be

$$f(x + h) = f(x) + hf'(x) + \frac{h^2}{2!}f''(x) + \frac{h^3}{3!}f'''(x) + \frac{h^4}{4!}f''''(x) + \cdots,$$

the form of the general term on the right-hand side being given by the formula $\dfrac{h^n}{n!}f^{(n)}(x)$, where $f^{(n)}(x)$ means the nth differential coefficient of $f(x)$.

This celebrated formula is known as Taylor's series, and its existence is usually referred to as Taylor's theorem. Brook Taylor was an English mathematician (1685–1731). His name has become inseparably attached to the formula, but it is doubtful whether he was actually the first to discover it. It is said to have been known by James Gregory (1638–1675). It is one of the curiosities of mathematical history that it was apparently not discovered by Newton, though Newton knew some particular cases of it. The reader who has found the foregoing sections rather sophisticated may take comfort from the thought that even Newton never quite got this point of view.

On putting $x = 0$ in Taylor's theorem, we obtain

$$f(h) = f(0) + hf'(0) + \frac{h^2}{2!}f''(0) + \cdots.$$

This is usually known as Maclaurin's theorem. It is not materially different from Taylor's theorem, since it is easy to get back from Maclaurin's theorem to Taylor's theorem merely by a change of notation. There are many striking particular cases of this formula. The binomial theorem, for example, is

$$(1 + x)^n = 1 + nx + \frac{n(n-1)}{1 \cdot 2}x^2 + \frac{n(n-1)(n-2)}{1 \cdot 2 \cdot 3}x^3 + \cdots$$

(here the x corresponds to the h in the previous formula). Then there are series which express $\sin x$ and $\cos x$ in powers of x; they are

$$\sin x = x - \frac{x^3}{3!} + \frac{x^5}{5!} - \frac{x^7}{7!} + \cdots$$

and

$$\cos x = 1 - \frac{x^2}{2!} + \frac{x^4}{4!} - \frac{x^6}{6!} + \cdots.$$

Another simple series, discovered by Gregory, is that for the inverse tangent. Suppose that x is defined as a function of y by the formula $x = \tan y$. Then conversely y is a function of x, and the relationship is expressed by the notation $y = \tan^{-1} x$. This function has the expansion

$$\tan^{-1} x = x - \frac{x^3}{3} + \frac{x^5}{5} - \frac{x^7}{7} + \cdots$$

and this is Gregory's series. The formula for π known as Gregory's formula is the particular case of this obtained by taking $x = 1$, $\tan^{-1} 1$ (the angle whose tangent is 1) being $\frac{1}{4}\pi$ radians.

The exponential function.

There are a number of mathematical problems which reduce to the question of finding a function which is equal to its own differential coefficient, for all values of the variable; in other words, we want to find a function $f(x)$ such that $f'(x) = f(x)$ for all values of x. We have not yet encountered such a function, and actually it is not to be found among the elementary functions such as x, x^2, or x^3.

To make headway with this problem we shall use a method which is often effective in mathematics. Suppose that the problem had been solved, and that we knew such a function. Then it would have to have certain properties and to satisfy certain formulae. From among these formulae we can perhaps point to one by which the function can actually be defined. Our hypothesis that it exists is thus confirmed, and the whole theory can then be fitted together on this basis.

This is what we shall do with the present problem. Suppose that a function $f(x)$ such that $f'(x) = f(x)$ exists, and suppose that it is the sort of function which can be expressed by means of Taylor's theorem. Now if $f'(x)$ is equal to $f(x)$, then $f''(x)$, the derivative of $f'(x)$, is also the derivative of $f(x)$, i.e., it is $f'(x)$, which is equal to $f(x)$. Hence the second derivative is equal to the original function; and similarly it can be proved in turn that all the derivatives are equal to the original function. Thus the Taylor's series

$$f(x+h) = f(x) + hf'(x) + \frac{h^2}{2!}f''(x) + \cdots$$

becomes in this case simply

$$f(x+h) = f(x) + hf(x) + \frac{h^2}{2!}f(x) + \cdots$$

or

$$f(x+h) = f(x)\left(1 + h + \frac{h^2}{2!} + \frac{h^3}{3!} + \cdots\right).$$

In the Maclaurin form, in which $x = 0$, this becomes

$$f(h) = f(0)\left(1 + h + \frac{h^2}{2!} + \frac{h^3}{3!} + \cdots\right).$$

Here $f(0)$ does not depend on h. It is just a constant, and there is nothing to say what its value is. We can choose it to be anything we like, and the simplest choice is to take it to be 1. The formula then reduces to

$$f(h) = 1 + h + \frac{h^2}{2!} + \frac{h^3}{3!} + \cdots.$$

Now it can be proved that the series on the right-hand side is convergent, whatever value h may have. Consequently it has a definite "sum." It is this sum which is taken to be the value of $f(h)$. This is our *definition.* Also it can be shown without much difficulty that the function defined in this way does actually solve the original problem. This is merely a matter of differentiating the above series term-by-term, and using the formula $\frac{d}{dh}h^n = nh^{n-1}$. On doing this it is found that the series simply reproduces itself. The function which has been defined in this way is known as the exponential function. Turning back to the Taylor formula, this is now seen to be equivalent to

$$1 + (x + h) + \frac{(x + h)^2}{2!} + \cdots$$

$$= \left(1 + x + \frac{x^2}{2!} + \cdots\right) \times \left(1 + h + \frac{h^2}{2!} + \cdots\right).$$

This can be verified by actual multiplication. The reader can at any rate do this for the terms which have actually been written down. For the later terms, a certain amount of algebra is required.

This formula has some interesting consequences. If we take x and h both equal to 1, it becomes

$$1 + 2 + \frac{2^2}{2!} + \cdots = \left(1 + 1 + \frac{1}{2!} + \cdots\right)^2$$

or, in the $f(\ )$ notation,

$$f(2) = \{f(1)\}^2.$$

If we take $x = 2$ and $h = 1$, it becomes

$$f(3) = f(2) \times f(1),$$

or, in view of the previous result,

$$f(3) = \{f(1)\}^3.$$

We can carry on with this argument indefinitely. It is clear that the number $f(1) = 1 + 1 + \frac{1}{2!} + \frac{1}{3!} + \cdots$ is of special importance in the theory, and a special notation is assigned to it. It is the celebrated number e. Some of the properties of this number have already been discussed. The result of the above argument can now be expressed by the formula

$$f(n) = e^n$$

where n is any positive integer.

It can be shown that this formula, proved here for integers, is also true if n is replaced by a fraction, or even by an irrational number. Consequently the exponential function $1 + x + \frac{x^2}{2!} + \cdots$ is usually denoted by e^x. The formula derived from Taylor's series can be written simply as

$$e^{x+h} = e^x \times e^h.$$

Chapter XIV

THE INTEGRAL CALCULUS

The general problem of area.

The problem of the area of a flat surface was solved in Chapter IX in the case of a rectangle, a triangle, or a circle. We shall now consider whether our ideas about area can be extended to other regions. This is a problem of some interest in itself, but it is also interesting as being one of the simplest examples of the subject known as the integral calculus.

Suppose that I take a piece of squared paper, divided up into little squares by the lines printed on it. I draw on this any closed curve, or closed figure of any kind with a boundary made up of straight lines or curves. If this boundary does not cross itself anywhere, it will be seen that it divides the paper into two parts, one inside and the other outside the boundary. The question is whether the region inside the boundary has a definite area.

In a general way, one might take as an approximate value of the area the number of little squares which lie entirely inside the boundary. There may be many squares along the boundary which lie partly inside and partly outside it, but as it is not clear how to count these, it is perhaps best to ignore them.

If the boundary was of a very complicated kind, or

if the figure drawn was very long and thin, there might be a great many of these squares partly inside and partly outside the region. The number of squares entirely inside would then not give a fair idea of the whole figure. We could however, in thought at any rate, divide the whole figure up again by a still finer network of lines. For example, the side of each square might be divided into ten equal parts, so that each square would be divided into a hundred smaller squares. The number of these smaller squares lying inside the original boundary (in proportion to the total number on the page) would generally give a better idea of its area.

Of course we are really thinking about a plane defined theoretically by number-pairs (x, y), and the boundary of our region is defined by one or more equations connecting x and y. It is not even obvious that such a boundary does divide the plane into two parts, an interior and an exterior. After much thought mathematicians have decided that any ordinary kind of boundary does divide up the plane in this way. The problem of the area of the interior can then be approached by the above method. In thought, of course, there is no barrier to the fineness of the mesh that can be used. As we make it finer and finer we may hope to get better and better approximations to the area. The actual area will be defined as the limit, if it exists, of these approximations.

Let us see whether this method works in the case of a triangle. Take for example a triangle in the Cartesian (x, y) plane with vertices at the points $(0, 0)$, $(a, 0)$, and (a, b). The sides of this triangle are the line $y = 0$, the line $x = a$, and the line $y = bx/a$. These together make up the boundary of a triangle of base a and height b.

The reader will no doubt say, "But I know the area of this triangle; it is $\frac{1}{2}ab$." The point however is that in mathematics a figure does not have an area, or any other property of this kind, just by nature. It has it according to certain definitions which we make. In the previous section in which this was discussed, $\frac{1}{2}ab$ was really the definition of the area. Now we are trying to

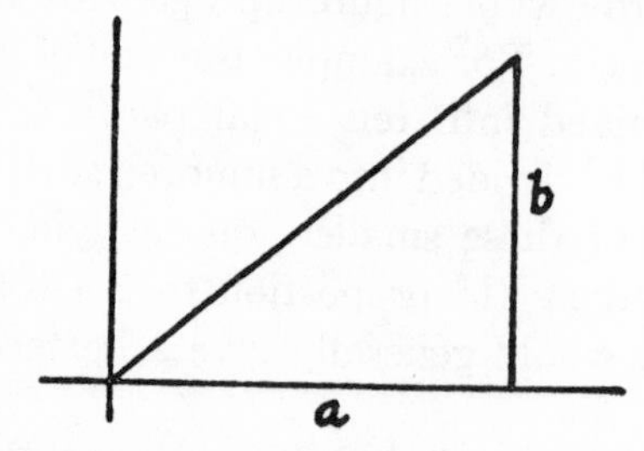

define area in a new way, and it is not quite obvious that the result will be the same as before. It is highly desirable that it should be, in the case of ordinary figures such as a triangle. But there might be very extraordinary figures, which would have an area according to one definition but not according to another.

To proceed, suppose that the (x, y) plane is divided up into a network of small squares of side h, by the lines $x = 0, h, 2h, 3h, \cdots$, and the lines $y = 0, h, 2h, 3h, \cdots$. Each square is of area h^2, and we have to count up how many of them lie inside the triangle. It is not easy to do this exactly, but we can do so sufficiently closely for our purpose. Suppose that h is just $1/n$th of the base a, so that n columns of squares have to be counted. The mth column lies between $x = (m - 1)h$ and $x = mh$, so that its left-hand side intersects the side of the triangle, whose equation is $y = bx/a$, where $y = b(m - 1)h/a$. Allowing for the fact that this may not be an exact multiple of h, we see that the number of squares in the column is at most $b(m - 1)/a$, and at least $b(m - 1)/a - 1$.

If the upper estimate, $b(m - 1)/a$, of the number of squares in the mth column were attained for every value of m, the total would be

$$\frac{b}{a}\left\{1 + 2 + \cdots + (n - 1)\right\} = \frac{b}{a}\,\tfrac{1}{2}\,(n - 1)n$$

by the rule for summing an arithmetical progression. The total area of these squares is

$$\frac{b}{a}\,\tfrac{1}{2}\,(n - 1)nh^2 = \tfrac{1}{2}\,\frac{b}{a}\,n^2h^2 - \tfrac{1}{2}\,\frac{b}{a}\,nh^2 = \tfrac{1}{2}ab - \tfrac{1}{2}bh$$

since $nh = a$. If we took the lower estimate every time, we should have to subtract an area $nh^2 = ah$. Whichever estimate we take, the result differs from $\frac{1}{2}ab$ by a term containing a factor h, which is therefore very small when h is very small. In other words, the limit of the area of the little squares contained in the triangle, when their side h tends to zero, is $\frac{1}{2}ab$. This is therefore the area of the triangle according to the new definition. Fortunately, it agrees with the area according to the old definition.

The integral calculus.

The whole point of all this, of course, is that the method applies to all sorts of figures, and not merely to triangles. In other cases, however, the details might be rather formidable, as we might have to evaluate sums much more complicated than arithmetical progressions. Let us see whether the process cannot be simplified.

In the first place, in the above example there is no particular point in counting up the exact number of squares in the mth column. We might just as well take instead the tallest rectangle between the lines $x = (m - 1)h$ and $x = mh$. The height of this rectangle is $b(m - 1)h/a$, and its area is therefore $b(m - 1)h^2/a$. It differs from the sum of the corresponding squares by at most a part of one square. The sum of all these terms is $\frac{1}{2}ab - \frac{1}{2}bh$ as before, so that we obtain the same result from this method as from the previous one.

Now suppose that we wanted to find the area, not just of an ordinary triangle, but of a triangle of which one side was a curve defined by an equation of the form $y = f(x)$; for example it might be $y = x^2$ or $y = x^3$.

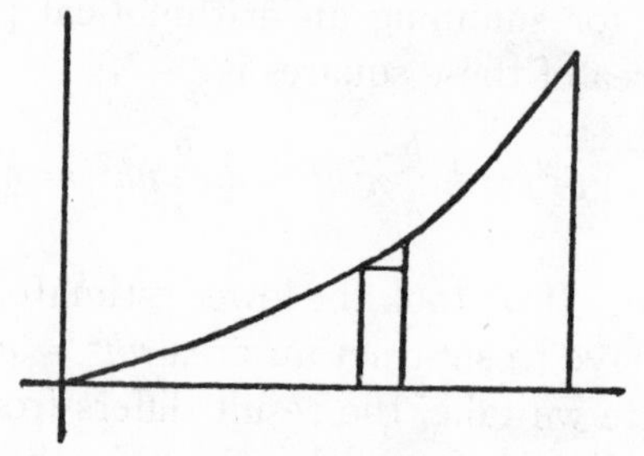

The procedure would be exactly the same. Divide up the range of variation of x into a large number of small parts by the lines $x = 0, h, 2h, \cdots$. If $f(x)$ is steadily increasing, the tallest rectangle which can be fitted into the figure between the lines $x = (m - 1)h$ and $x = mh$ is of height $f\{(m - 1)h\}$. Its area is this multiplied by h. The area to be assigned to the whole region is therefore the limit of the sum of all the terms $f\{(m - 1)h\} \times h$, when h is very small. The smaller h is, of course, the larger the number of terms will be, but the smaller each individual term will be.

A special notation has come into use for the limit of such sums. We think of the typical slice cut out of the area as beginning at x, and ending at $x + \delta x$, so that δx corresponds to the h of the above argument. (As in differential calculus, δx here just means "a small addition to x," not δ times x.) The typical term of the above sum is then $f(x)\delta x$. The sum of all such terms is then denoted by $\sum f(x)\delta x$. The limiting value of this sum when the number of terms becomes very large, and the individual δx very small, is denoted by $\int f(x)dx$. The δ has turned into d, and the $\sum$ into the old-fashioned long s. This is a purely conventional expression, and it must not be thought that it can be taken to pieces, and a separate meaning assigned to each piece. The whole thing is known as an integral, and the science of such processes is called the integral calculus.

$\int$ is always read as "integral," and $\int f(x)dx$ as "the integral of $f(x)$ with respect to x," or just as "integral $f(x)dx$." The process of finding the value of an integral is called integration. Actually, the expression is usually elaborated a little. In the above example, x was supposed to vary between 0 and a. These are called the limits of integration, or the lower and upper limits, and are put into the formula as follows: $\int_0^a f(x)dx$. In the case of the above triangle with straight line sides, the function $f(x)$ is $\frac{bx}{a}$, and the result of our investigation of the area of the triangle would be written as

$$\int_0^a \frac{bx}{a}\,dx = \tfrac{1}{2}ab.$$

In most cases it is not easy to evaluate integrals directly. The only very easy case is that in which the function $f(x)$ is a constant, say C. Suppose that x varies between the limits a and b. The typical term of the sum $\sum f(x)\delta x$ is just $C\delta x$. Here δx is the breadth of the typical slice, and the sum of all these breadths is $b - a$. The value of the sum is therefore $C(b - a)$, and the integral has the same value. Thus

$$\int_a^b Cdx = C(b - a).$$

The corresponding geometrical problem is simply that of finding the area of a rectangle of base $b - a$ and height C.

These problems about areas are typical problems of the integral calculus, but there are many others. What is the length of a given curve? What is the volume of a solid body of given shape? What is the area of its curved surface? How far will a body, moving with a given law of velocity, go in a given time? All such problems the integral calculus sets out to solve.

In Chapter IX we obtained the area of a circle, and the length of its circumference, by methods slightly different from those of this chapter. The previous methods were really examples of integration too; they depended on the same essential idea, that of finding the limit of a very large number of very small terms. The only difference was that, in the problem of the area of a circle, we divided up the figure into a large number of little triangles instead of little squares or strips, because the triangles happened to fit into the circle in a simpler way than squares would.

Differentiation of the integral.

So far, the differential calculus and the integral calculus have appeared to be quite independent subjects. Actually there is a close connection between them.

Consider the formula for the area of a triangle with one curved side $y = f(x)$, the base going from $x = 0$ to $x = a$. In the notation of the integral calculus, this area is $\int_0^a f(x)dx$. In forming the integral we supposed that a was a fixed number; but if you think of the same thing being done with different values of a, the result will depend on the value of a. In other words, the integral is a function of a. We might denote it by $F(a)$, so that

$$F(a) = \int_0^a f(x)dx.$$

Now the differential calculus expert, on seeing a function, naturally wants to differentiate it. What is the differential coefficient of $F(a)$?

The rule for differentiation is, change a into $a + h$, take the difference $F(a + h) - F(a)$, divide by h, and proceed to the limit when h tends to zero. In this case $F(a + h) - F(a)$ is the area between the curved boundary, the x-axis, and the lines $x = a$ and $x = a + h$. If h is very small, it will be seen, on drawing a reasonable figure, that this area is approximately $f(a) \times h$. The result of dividing by h and proceeding to the limit is therefore simply $f(a)$. Hence $\frac{d}{da} F(a) = f(a)$; that is, the differential coefficient of the integral is the value, at the upper limit of integration, of the function integrated. One can say roughly that differentiation is the inverse process to integration.

Differentiation of the integral gets you back to the original function. This is a very important discovery, because it is usually very much easier to do differentiation than integration. In the differential calculus one can accumulate a large stock of formulae, giving the differential coefficients of all sorts of functions. Suppose that this has been done, and that we wish to evaluate the integral $F(a) = \int_0^a f(x)dx$. We search among our stock of formulae for a function whose differential coefficient is equal to $f(x)$ (I mean of course for all relevant values of x). Suppose that we find one, and that it is $\phi(x)$; that is, $\frac{d}{dx} \phi(x) = f(x)$. Now we also know that $\frac{d}{dx} F(x) = f(x)$. Consequently $\frac{d}{dx} (F(x) - \phi(x)) = f(x) - f(x) = 0$, and hence (see p. 164), $F(x) - \phi(x)$ must be a constant. That is, there is a constant C such that $f(x) = \phi(x) + C$ for all values of x. The value of the constant C can usually be found by giving x some particular value. For example, in the figure which we have considered above, $F(x)$ is clearly 0 when x is 0, and consequently C is minus the value of $\phi(x)$ when

$x = 0$, or $-\phi(0)$. The result could then be written in the form

$$\int_0^a f(x)dx = \phi(a) - \phi(0).$$

All this can be illustrated by investigating still once more the problem of the area of a triangle. Suppose that the inclined boundary of the triangle is the line $y = kx$. It is then a question of finding a function of which this is the differential coefficient. Even the very small stock of functions differentiated in this book contains the answer to this; it is $\frac{1}{2}kx^2$ (see p. 153). If $F(a) = \int_0^a kxdx$, it follows that $F(a)$ differs from $\frac{1}{2}ka^2$ by a constant, that is a number independent of a. Since both $F(a)$ and $\frac{1}{2}ka^2$ vanish when $a = 0$, this constant is in fact 0. The result is therefore

$$\int_0^a kxdx = \tfrac{1}{2}ka^2.$$

This agrees with the result previously obtained for the area of a triangle, as we see on putting $k = b/a$.

It might happen that there was no function among our stock which had the required differential coefficient, and then we should not be so fortunate. This would not mean that the problem of integration was insoluble, but merely that its solution was not of any form already known to us. We should have to try to enlarge our knowledge of functions. Such situations have led to many advances in mathematical science.

The integral calculus would not have been entirely strange to the mathematicians of antiquity. They invented ingenious methods of calculating certain areas and volumes, and their methods might be regarded as a sort of integration. But as they did not know the differential calculus, they never found an easy and systematic way of evaluating integrals.

Logarithms to the base e.

We have shown that, in mathematics which is used for the purpose of actual calculations, logarithms to the base 10 are always used. But there are many branches of mathematics in which it is the structure of the subject, the way the formulae fit together, with which we are primarily concerned, and in which actual calculations, if they occur at all, are a secondary consideration. In this sort of mathematics the number e is always used as the base of logarithms. We shall now try to explain why it is this number which is used.

One of the principal reasons is that, in the integral calculus, many problems involve the integration of the function $1/x$; and the integral of this function is $\log_e x$. The number e arises naturally, so to speak, when we take this point of view.

To prove that the integral of $1/x$ is $\log_e x$ is equivalent to proving that the differential coefficient of $\log_e x$ is $1/x$; or, if we write $y = \log_e x$, we have to prove that $\frac{dy}{dx} = \frac{1}{x}$. Now the relation $y = \log_e x$ is equivalent to $x = e^y$; we have shown in the previous chapter that this relation gives $\frac{dx}{dy} = e^y$, or, what is the same thing, $\frac{dx}{dy} = x$. Now $\frac{dy}{dx} = \frac{1}{\frac{dx}{dy}}$ (this is a consequence of the relation $\frac{\delta y}{\delta x} = \frac{1}{\frac{\delta x}{\delta y}}$ in which the symbols just mean fractions); hence $\frac{dy}{dx} = \frac{1}{x}$, the required result.

We can put this in a slightly different way. Consider the problem of evaluating the integral $\int_1^a \frac{dx}{x}$, where a

is any number greater than 1. According to the general method of the integral calculus, this integral will only differ by a constant from any function $\phi(a)$ whose differential coefficient is $1/a$. Such a function is $\log_e a$, and therefore

$$\int_1^a \frac{dx}{x} = \log_e a + C,$$

where C is a constant, i.e., is independent of a. But on putting $a = 1$, the integral on the left-hand side becomes 0, and $\log_e 1$ is also 0. Hence C is 0, and the result is simply

$$\int_1^a \frac{dx}{x} = \log_e a.$$

This formula, which is made up from very simple elements, again demonstrates the occurrence of e in a natural way.

Some writers have taken this formula as the *definition* of a logarithm. If this is made the starting-point, the whole theory is reversed, and the exponential function appears as the inverse of the logarithmic function, but the final result is the same as before.

We could, of course, operate always with logarithms to the base 10 if we insisted on doing so, but then most of our formulae would involve a certain constant. This arises from the formula

$$\log_{10} x = \mu \log_e x,$$

where x is any positive number, and μ is $1/\log_e 10$, and is equal to .43429 to five decimal places. For example, the above formula of the integral calculus would have to be written as

$$\int_1^a \frac{dx}{x} = \frac{1}{\mu} \log_{10} a.$$

Of course if we have obtained any formula involving

logarithms to the base e, and then have to evaluate it numerically by means of ordinary log tables to base 10, we can do so at once by using the same formula. Suppose for example that it is required to evaluate the integral $\int_1^3 \frac{dx}{x}$. This is equal to $\log_e 3$, and so to $\frac{1}{\mu}\log_{10} 3$. From the tables we find that $\log_{10} 3 = .4771$, and dividing this by μ the result is about 1.1.

I will conclude by giving two more examples of the occurrence of logarithms to the base e in a natural way, in formulae which have no primary connection with the integral or differential calculus.

The first example concerns the series $1 + \frac{1}{2} + \frac{1}{3} + \frac{1}{4} + \cdots$. We proved in Chapter VIII that this series is divergent, i.e., that the partial sums $1 + \frac{1}{2} + \frac{1}{3} + \cdots + \frac{1}{n}$ become indefinitely large as n is increased. It is then natural to ask if the sum can be compared, as regards its rate of growth as n tends to infinity, with any simple function of n. Does it behave in approximately the same way as n, or as n^2, or as $\sqrt{n}$, or what? The answer is that it behaves in approximately the same way as $\log_e n$. The relation is written in symbols as

$$1 + \tfrac{1}{2} + \tfrac{1}{3} + \cdots + \frac{1}{n} \sim \log_e n.$$

This is read as "the expression on the left-hand side is asymptotically equivalent to $\log_e n$." All that it means is that the left-hand side divided by the right-hand side (in this case $\log_e n$) tends to the limit 1 as n tends to infinity. The proof of this theorem is too technical to be given here, though actually it is not at all difficult.

The expression "asymptotically" comes from geometry. An asymptote (from Greek "not intersecting") is a straight line towards which a curve approaches indefinitely closely, without ever actually meeting it. One of the simplest examples is the curve $y = 1/x$, which, as x tends to infinity, approaches the line $y = 0$ in this way; (the reader should draw a figure). Thus we say that $y = 0$ is an asymptote of the curve $y = 1/x$.

The prime-number theorem.

Our second example deserves a head-line of its own, since it belongs to a very different kind of mathematics from the previous one. We proved in Chapter II the famous theorem of Euclid that there is an infinity of prime numbers; that is that, however far we go along the numbers, there must always be more prime numbers beyond the point at which we have arrived. This theorem remained practically all that was known about the distribution of prime numbers until quite modern times. But then mathematicians began to ask more searching questions about the problem. These enquiries took the form of asking, about how many prime numbers are there not exceeding x, when x is large, say x is a thousand, or a million, or a million million?

The number of prime numbers not exceeding x is denoted by the symbol $\pi(x)$. This $\pi(\;)$, which no doubt arose from the initial letter of "prime," has nothing to do with the constant π, but is a functional symbol meaning what has just been explained. For example, the prime numbers less than 10 are 2, 3, 5, and 7, so that $\pi(10) = 4$. Up to 17, we have also the primes, 11, 13 and 17, so that $\pi(17) = 7$, and so on. The question then is whether there is an asymptotic formula for $\pi(x)$, as x tends to infinity, in the same sense as the asymptotic formula for the sum $1 + \frac{1}{2} + \cdots + \frac{1}{n}$ which has just been explained. It is not obvious that there is such a formula, because the prime numbers occur in a rather irregular way. The problem was the subject of many researches during the nineteenth century, until at last in 1896 it was solved independently by two mathematicians, Hadamard in France and de la Vallée-Poussin in Belgium. The result is

$$\pi(x) \sim \frac{x}{\log_e x},$$

i.e., when x is large, $\pi(x)$ is approximately x divided by the logarithm of x to the base e.

This is the prime-number theorem. Again logarithms to the base e occur in a problem which in the first place involves only the simplest elements of mathematics, the integers. It would be quite impossible to give here any idea of how the prime-number theorem was proved.

Chapter XV

AFTERMATH

Bertrand Russell said that mathematics is the science in which we do not know what we are talking about, and do not care whether what we say about it is true. This paradox means in the first place that mathematicians are people who put together patterns of certain kinds. The patterns must be made up of something, but the ordinary mathematician does not usually concern himself about what the something really is. It may be different things in different cases or to different people, but whether it is so is a question for philosophers. As mathematicians we do not know what it is that we are talking about. As to not caring whether what we say is true, perhaps this means that many different kinds of primary axioms could form the starting points of mathematical systems. The mathematicians would only be concerned to follow out their consequences, not to enquire about the comparative validity of different sets of axioms.

In this book we have given elementary introductions to various branches of mathematics, but no attempt at a complete survey has been made. Algebra and geometry are to form the subjects of further volumes of this series, so that very little has been said about them here. Dynamics and subjects of that kind, usually known as applied mathematics, have only been mentioned casually. The main subject which we have dealt with is what mathematicians call analysis. This is a rather vague expression for those parts of mathematics in which the ideas of limit, variation, function, and so on, are uppermost. The experts in these subjects sometimes describe themselves as analysts. An analyst should be able to handle such things as integrals and infinite series just as well as if they were the simple expressions of elementary algebra. The expression "$+ \cdots$" is not uncommonly used in mathematical writings to mean something which the writer proposes to ignore, in the hope that it does not really matter very much. Analysts also use this expression, but they should know, each time they use it, exactly what they mean by it.

Mathematics is a highly technical subject. If you take down a book from a mathematician's shelves and open it at random, it is very likely that on the first page which you read, you will not be able to understand anything at all. Still I hope that anyone who has read this book would feel that, even if he was reading a foreign language, it was not written in an unknown script. He might be able to form some idea of the sort of thing that was going on, even if he could not actually follow the details of the working. At present even a mathematician cannot usually follow the writings of other mathematicians without a special study of their particular subjects. The time when any one person could know the whole of mathematics is long past. The accumulated stock of mathematical knowledge is very large, and is still growing rapidly. All that any mathematician can do now is to concentrate on those topics which he finds specially interesting. In this way it is possible to reach the limits of knowledge on fairly narrow fronts, and to make progress there, while remaining comparatively ignorant of other parts of the subject.

It is impossible in a book of this kind to teach mathematical technique. There is no short cut to this. Anyone who wants to be able to solve mathematical problems must go through the ordinary routine. This is what mathematical text books are for. There are many good ones at the present time. Much of the fascination of mathematics lies in the scope it gives for the use of complicated techniques. One has to take trouble to learn to use them, but most people who have done so seem to have found it well worth while.

I find that many people, even those working in other branches of learning, do not know whether mathematics is, like science, active at the present day, or whether it is merely a routine which has come down to us from time immemorial. Actually, I suppose that more people are engaged in mathematical research now than at any previous time. Some topics become exhausted, of course; so far as I know, no one now discovers anything new about trigonometry, for example. At the same time new subjects open out before us, so that there always seems to be plenty to do.

Progress has been continuous for several hundred years now, and shows no signs of slackening. Probably, as long as there are mathematicians, some of them will be finding out something new about their subjects.

The question is sometimes asked, what have mathematicians discovered in modern times which would have been completely new and strange to the Greeks? One of the best answers to this is, the theory of functions of a complex variable. This is a subject which we have not been able to touch on here, but which has occupied a very large part of the time of mathematicians during the last century and a half. Briefly, this is what it is about. We have introduced here the idea of a function, and the idea of a complex number. Now put the two ideas together. We can define functions in which the independent variable is not a "real number" but a "complex number" (x, y), or $x + iy$, in the sense of Chapter X. Such a number (x, y) or $x + iy$ is usually denoted simply by z. Then in formulae such as x^2 or x^3 we can just replace the x by z and think instead about z^2 or z^3. In this way we are led to consider functions of a complex variable. The theory of such functions contains many very remarkable theorems, particularly those due to the great French mathematician Cauchy (1789–1857). Cauchy's theory of functions of a complex variable would have surprised the Greeks very much, and surely it would have delighted them too.

Perhaps the most surprising thing about mathematics is that it is so surprising. The rules which we make up at the beginning seem ordinary and inevitable, but it is impossible to foresee their consequences. These have only been found out by long study, extending over many centuries. Much of our knowledge is due to a comparatively few great mathematicians such as Newton, Euler, Gauss, Cauchy or Riemann; few careers can have been more satisfying than theirs. They have contributed something to human thought even more lasting than great literature, since it is independent of language.

It is sometimes supposed that mathematicians have extraordinarily remote and mysterious minds, or that they are people who can think quite easily about the inconceivable. This is not so. Some of the patterns which they make up are extremely complicated, so that it almost passes the power of the human mind to see whether they fit together correctly or not. But essentially their patterns are of the same sort as the simple ones of which we have given some examples here.

Mathematicians are often asked why they spend their lives trying to solve such curious problems. What good is it to know that every number is the sum of four squares? Why do you want to know about prime-pairs? What does it matter whether π is rational or irrational?

A mathematician faced with these questions is in much the same position as a composer of music being questioned by someone with no ear for music. Why do you select some sets of notes and have them repeated by musicians, and reject others as worthless? It is difficult to answer except to say that there are harmonies in these things which we find that we can enjoy. It is true of course that some mathematics is useful. The invention of logarithms was welcomed by astronomers because it reduced the labour of their calculations. The theory of differential equations enables engineers to think about such things as the flow of water in pipes. The theory of linear operators enables the physicist to think about the atom. But the so-called pure mathematicians do not do mathematics for such reasons. It can be of no practical use to know that π is irrational, but if we can know, it would surely be intolerable not to know. Pure mathematicians do mathematics because it gives them an aesthetic satisfaction which they can share with other mathematicians. They do it because for them it is fun, in the same way perhaps that people climb mountains for fun. It may be an extremely arduous and even fatal pursuit, but it is fun nevertheless. Mathematicians enjoy themselves because they do sometimes get to the top of their mountains, and anyhow trying to get up does seem to be worth while.

I once heard a lecture by a physicist in which he derided what he thought were the futilities of pure mathematics; but then he referred to some theorem of pure mathematics which, fifty years after its discovery, had found an application in relativity, and this seemed to him little short of miraculous. But such cases are not uncommon. The ellipse was studied for centuries before it was found to be the orbit of a planet. To express astonishment at this is to mistake the nature of mathematics. Mathematicians are engaged in discovering and mapping out a real world. It is a world of thought, but it is of a kind on the basis of which the physical world is, to a certain extent, also constructed.

BIBLIOGRAPHY

M. BLACK: *The Nature of Mathematics.* (Kegan Paul, Trench, Trubner & Co., Ltd.)

F. CAJORI: *A History of Mathematics.* (Macmillan.)

R. COURANT and H. ROBBINS: *What is Mathematics?* (O.U.P.)

C. V. DURELL and R. M. WRIGHT: *Elementary Trigonometry.* (G. Bell & Sons, Ltd.)

R. C. FAWDRY and C. V. DURELL: *Calculus for Schools.* (Edward Arnold & Co.)

W. L. FERRAR: *Higher Algebra for Schools.* (Clarendon Press.)

G. H. HARDY: *A Mathematician's Apology.* (C.U.P.)

G. H. HARDY: *A Course of Pure Mathematics.* (C.U.P.)

G. H. HARDY and E. M. WRIGHT: *An Introduction to the Theory of Numbers.* (Clarendon Press.)

BERTRAND RUSSELL: *Introduction to Mathematical Philosophy.* (George Allen and Unwin, Ltd.)

D. M. Y. SOMMERVILLE: *Analytical Conics.* (C.U.P.)

A. N. WHITEHEAD: *An Introduction to Mathematics.* (Home University Library, O.U.P.)

A CATALOGUE OF SELECTED DOVER BOOKS IN ALL FIELDS OF INTEREST

A CATALOGUE OF SELECTED DOVER BOOKS IN ALL FIELDS OF INTEREST

DRAWINGS OF WILLIAM BLAKE, William Blake. 92 plates from Book of Job, *Divine Comedy, Paradise Lost,* visionary heads, mythological figures, Laocoon, etc. Selection, introduction, commentary by Sir Geoffrey Keynes. 178pp. 8⅛ x 11. 22303-5 Pa. $5.00

ENGRAVINGS OF HOGARTH, William Hogarth. 101 of Hogarth's greatest works: *Rake's Progress, Harlot's Progress, Illustrations for Hudibras, Before and After, Beer Street and Gin Lane,* many more. Full commentary. 256pp. 11 x 13¾. 22479-1 Pa. $12.95

DAUMIER: 120 GREAT LITHOGRAPHS, Honore Daumier. Wide-ranging collection of lithographs by the greatest caricaturist of the 19th century. Concentrates on eternally popular series on lawyers, on married life, on liberated women, etc. Selection, introduction, and notes on plates by Charles F. Ramus. Total of 158pp. 9⅜ x 12¼. 23512-2 Pa. $6.00

DRAWINGS OF MUCHA, Alphonse Maria Mucha. Work reveals draftsman of highest caliber: studies for famous posters and paintings, renderings for book illustrations and ads, etc. 70 works, 9 in color; including 6 items not drawings. Introduction. List of illustrations. 72pp. 9⅜ x 12¼. (Available in U.S. only) 23672-2 Pa. $4.50

GIOVANNI BATTISTA PIRANESI: DRAWINGS IN THE PIERPONT MORGAN LIBRARY, Giovanni Battista Piranesi. For first time ever all of Morgan Library's collection, world's largest. 167 illustrations of rare Piranesi drawings—archeological, architectural, decorative and visionary. Essay, detailed list of drawings, chronology, captions. Edited by Felice Stampfle. 144pp. 9⅜ x 12¼. 23714-1 Pa. $7.50

NEW YORK ETCHINGS (1905-1949), John Sloan. All of important American artist's N.Y. life etchings. 67 works include some of his best art; also lively historical record—Greenwich Village, tenement scenes. Edited by Sloan's widow. Introduction and captions. 79pp. 8⅜ x 11¼. 23651-X Pa. $5.00

CHINESE PAINTING AND CALLIGRAPHY: A PICTORIAL SURVEY, Wan-go Weng. 69 fine examples from John M. Crawford's matchless private collection: landscapes, birds, flowers, human figures, etc., plus calligraphy. Every basic form included: hanging scrolls, handscrolls, album leaves, fans, etc. 109 illustrations. Introduction. Captions. 192pp. 8⅞ x 11¾. 23707-9 Pa. $7.95

DRAWINGS OF REMBRANDT, edited by Seymour Slive. Updated Lippmann, Hofstede de Groot edition, with definitive scholarly apparatus. All portraits, biblical sketches, landscapes, nudes, Oriental figures, classical studies, together with selection of work by followers. 550 illustrations. Total of 630pp. 9⅛ x 12¼. 21485-0, 21486-9 Pa., Two-vol. set $17.90

THE DISASTERS OF WAR, Francisco Goya. 83 etchings record horrors of Napoleonic wars in Spain and war in general. Reprint of 1st edition, plus 3 additional plates. Introduction by Philip Hofer. 97pp. 9⅜ x 8¼. 21872-4 Pa. $4.50

THE EARLY WORK OF AUBREY BEARDSLEY, Aubrey Beardsley. 157 plates, 2 in color: *Manon Lescaut, Madame Bovary, Morte Darthur, Salome,* other. Introduction by H. Marillier. 182pp. 8⅛ x 11. 21816-3 Pa. $6.50

THE LATER WORK OF AUBREY BEARDSLEY, Aubrey Beardsley. Exotic masterpieces of full maturity: *Venus and Tannhauser, Lysistrata, Rape of the Lock, Volpone,* Savoy material, etc. 174 plates, 2 in color. 186pp. 8⅛ x 11. 21817-1 Pa. $5.95

THOMAS NAST'S CHRISTMAS DRAWINGS, Thomas Nast. Almost all Christmas drawings by creator of image of Santa Claus as we know it, and one of America's foremost illustrators and political cartoonists. 66 illustrations. 3 illustrations in color on covers. 96pp. 8⅜ x 11¼. 23660-9 Pa. $3.50

THE DORÉ ILLUSTRATIONS FOR DANTE'S DIVINE COMEDY, Gustave Doré. All 135 plates from Inferno, Purgatory, Paradise; fantastic tortures, infernal landscapes, celestial wonders. Each plate with appropriate (translated) verses. 141pp. 9 x 12. 23231-X Pa. $5.00

DORÉ'S ILLUSTRATIONS FOR RABELAIS, Gustave Doré. 252 striking illustrations of *Gargantua and Pantagruel* books by foremost 19th-century illustrator. Including 60 plates, 192 delightful smaller illustrations. 153pp. 9 x 12. 23656-0 Pa. $6.00

LONDON: A PILGRIMAGE, Gustave Doré, Blanchard Jerrold. Squalor, riches, misery, beauty of mid-Victorian metropolis; 55 wonderful plates, 125 other illustrations, full social, cultural text by Jerrold. 191pp. of text. 9⅜ x 12¼. 22306-X Pa. $7.00

THE RIME OF THE ANCIENT MARINER, Gustave Doré, S. T. Coleridge. Dore's finest work, 34 plates capture moods, subtleties of poem. Full text. Introduction by Millicent Rose. 77pp. 9¼ x 12. 22305-1 Pa. $4.50

THE DORE BIBLE ILLUSTRATIONS, Gustave Doré. All wonderful, detailed plates: Adam and Eve, Flood, Babylon, Life of Jesus, etc. Brief King James text with each plate. Introduction by Millicent Rose. 241 plates. 241pp. 9 x 12. 23004-X Pa. $6.95

THE COMPLETE ENGRAVINGS, ETCHINGS AND DRYPOINTS OF ALBRECHT DURER. "Knight, Death and Devil"; "Melencolia," and more—all Dürer's known works in all three media, including 6 works formerly attributed to him. 120 plates. 235pp. 8⅜ x 11¼. 22851-7 Pa. $7.50

MECHANICK EXERCISES ON THE WHOLE ART OF PRINTING, Joseph Moxon. First complete book (1683-4) ever written about typography, a compendium of everything known about printing at the latter part of 17th century. Reprint of 2nd (1962) Oxford Univ. Press edition. 74 illustrations. Total of 550pp. 6⅛ x 9¼. 23617-X Pa. $7.95

CATALOGUE OF DOVER BOOKS

THE COMPLETE WOODCUTS OF ALBRECHT DURER, edited by Dr. W. Kurth. 346 in all: "Old Testament," "St. Jerome," "Passion," "Life of Virgin," Apocalypse," many others. Introduction by Campbell Dodgson. 285pp. 8½ x 12¼. 21097-9 Pa. $7.50

DRAWINGS OF ALBRECHT DURER, edited by Heinrich Wolfflin. 81 plates show development from youth to full style. Many favorites; many new. Introduction by Alfred Werner. 96pp. 8⅛ x 11. 22352-3 Pa. $6.00

THE HUMAN FIGURE, Albrecht Dürer. Experiments in various techniques—stereometric, progressive proportional, and others. Also life studies that rank among finest ever done. Complete reprinting of *Dresden Sketchbook*. 170 plates. 355pp. 8⅜ x 11¼. 21042-1 Pa. $7.95

OF THE JUST SHAPING OF LETTERS, Albrecht Dürer. Renaissance artist explains design of Roman majuscules by geometry, also Gothic lower and capitals. Grolier Club edition. 43pp. 7⅞ x 10¾ 21306-4 Pa. $3.00

TEN BOOKS ON ARCHITECTURE, Vitruvius. The most important book ever written on architecture. Early Roman aesthetics, technology, classical orders, site selection, all other aspects. Stands behind everything since. Morgan translation. 331pp. 5⅜ x 8½. 20645-9 Pa. $5.00

THE FOUR BOOKS OF ARCHITECTURE, Andrea Palladio. 16th-century classic responsible for Palladian movement and style. Covers classical architectural remains, Renaissance revivals, classical orders, etc. 1738 Ware English edition. Introduction by A. Placzek. 216 plates. 110pp. of text. 9½ x 12¾. 21308-0 Pa. $10.00

HORIZONS, Norman Bel Geddes. Great industrialist stage designer, "father of streamlining," on application of aesthetics to transportation, amusement, architecture, etc. 1932 prophetic account; function, theory, specific projects. 222 illustrations. 312pp. 7⅞ x 10¾. 23514-9 Pa. $6.95

FRANK LLOYD WRIGHT'S FALLINGWATER, Donald Hoffmann. Full, illustrated story of conception and building of Wright's masterwork at Bear Run, Pa. 100 photographs of site, construction, and details of completed structure. 112pp. 9¼ x 10. 23671-4 Pa. $5.95

THE ELEMENTS OF DRAWING, John Ruskin. Timeless classic by great Viltorian; starts with basic ideas, works through more difficult. Many practical exercises. 48 illustrations. Introduction by Lawrence Campbell. 228pp. 5⅜ x 8½. 22730-8 Pa. $3.75

GIST OF ART, John Sloan. Greatest modern American teacher, Art Students League, offers innumerable hints, instructions, guided comments to help you in painting. Not a formal course. 46 illustrations. Introduction by Helen Sloan. 200pp. 5⅜ x 8½. 23435-5 Pa. $4.00

THE ANATOMY OF THE HORSE, George Stubbs. Often considered the great masterpiece of animal anatomy. Full reproduction of 1766 edition, plus prospectus; original text and modernized text. 36 plates. Introduction by Eleanor Garvey. 121pp. 11 x 14¾. 23402-9 Pa. $8.95

BRIDGMAN'S LIFE DRAWING, George B. Bridgman. More than 500 illustrative drawings and text teach you to abstract the body into its major masses, use light and shade, proportion; as well as specific areas of anatomy, of which Bridgman is master. 192pp. 6½ x 9¼. (Available in U.S. only) 22710-3 Pa. $4.50

ART NOUVEAU DESIGNS IN COLOR, Alphonse Mucha, Maurice Verneuil, Georges Auriol. Full-color reproduction of *Combinaisons ornementales* (c. 1900) by Art Nouveau masters. Floral, animal, geometric, interlacings, swashes—borders, frames, spots—all incredibly beautiful. 60 plates, hundreds of designs. 9⅜ x 8-1/16. 22885-1 Pa. $4.50

FULL-COLOR FLORAL DESIGNS IN THE ART NOUVEAU STYLE, E. A. Seguy. 166 motifs, on 40 plates, from *Les fleurs et leurs applications decoratives* (1902): borders, circular designs, repeats, allovers, "spots." All in authentic Art Nouveau colors. 48pp. 9⅜ x 12¼. 23439-8 Pa. $5.00

A DIDEROT PICTORIAL ENCYCLOPEDIA OF TRADES AND INDUSTRY, edited by Charles C. Gillispie. 485 most interesting plates from the great French Encyclopedia of the 18th century show hundreds of working figures, artifacts, process, land and cityscapes; glassmaking, papermaking, metal extraction, construction, weaving, making furniture, clothing, wigs, dozens of other activities. Plates fully explained. 920pp. 9 x 12. 22284-5, 22285-3 Clothbd., Two-vol. set $40.00

HANDBOOK OF EARLY ADVERTISING ART, Clarence P. Hornung. Largest collection of copyright-free early and antique advertising art ever compiled. Over 6,000 illustrations, from Franklin's time to the 1890's for special effects, novelty. Valuable source, almost inexhaustible.
Pictorial Volume. Agriculture, the zodiac, animals, autos, birds, Christmas, fire engines, flowers, trees, musical instruments, ships, games and sports, much more. Arranged by subject matter and use. 237 plates. 288pp. 9 x 12. 20122-8 Clothbd. $15.00

Typographical Volume. Roman and Gothic faces ranging from 10 point to 300 point, "Barnum," German and Old English faces, script, logotypes, scrolls and flourishes, 1115 ornamental initials, 67 complete alphabets, more. 310 plates. 320pp. 9 x 12. 20123-6 Clothbd. $15.00

CALLIGRAPHY (CALLIGRAPHIA LATINA), J. G. Schwandner. High point of 18th-century ornamental calligraphy. Very ornate initials, scrolls, borders, cherubs, birds, lettered examples. 172pp. 9 x 13. 20475-8 Pa. $7.95

ART FORMS IN NATURE, Ernst Haeckel. Multitude of strangely beautiful natural forms: Radiolaria, Foraminifera, jellyfishes, fungi, turtles, bats, etc. All 100 plates of the 19th-century evolutionist's *Kunstformen der Natur* (1904). 100pp. 9⅜ x 12¼. 22987-4 Pa. $5.00

CHILDREN: A PICTORIAL ARCHIVE FROM NINETEENTH-CENTURY SOURCES, edited by Carol Belanger Grafton. 242 rare, copyright-free wood engravings for artists and designers. Widest such selection available. All illustrations in line. 119pp. 8⅜ x 11¼. 23694-3 Pa. $4.00

WOMEN: A PICTORIAL ARCHIVE FROM NINETEENTH-CENTURY SOURCES, edited by Jim Harter. 391 copyright-free wood engravings for artists and designers selected from rare periodicals. Most extensive such collection available. All illustrations in line. 128pp. 9 x 12. 23703-6 Pa. $4.95

ARABIC ART IN COLOR, Prisse d'Avennes. From the greatest ornamentalists of all time—50 plates in color, rarely seen outside the Near East, rich in suggestion and stimulus. Includes 4 plates on covers. 46pp. 9⅜ x 12¼. 23658-7 Pa. $6.00

AUTHENTIC ALGERIAN CARPET DESIGNS AND MOTIFS, edited by June Beveridge. Algerian carpets are world famous. Dozens of geometrical motifs are charted on grids, color-coded, for weavers, needleworkers, craftsmen, designers. 53 illustrations plus 4 in color. 48pp. 8¼ x 11. (Available in U.S. only) 23650-1 Pa. $1.75

DICTIONARY OF AMERICAN PORTRAITS, edited by Hayward and Blanche Cirker. 4000 important Americans, earliest times to 1905, mostly in clear line. Politicians, writers, soldiers, scientists, inventors, industrialists, Indians, Blacks, women, outlaws, etc. Identificatory information. 756pp. 9¼ x 12¾. 21823-6 Clothbd. $65.00

HOW THE OTHER HALF LIVES, Jacob A. Riis. Journalistic record of filth, degradation, upward drive in New York immigrant slums, shops, around 1900. New edition includes 100 original Riis photos, monuments of early photography. 233pp. 10 x 7⅞. 22012-5 Pa. $7.00

NEW YORK IN THE THIRTIES, Berenice Abbott. Noted photographer's fascinating study of city shows new buildings that have become famous and old sights that have disappeared forever. Insightful commentary. 97 photographs. 97pp. 11⅜ x 10. 22967-X Pa. $6.00

MEN AT WORK, Lewis W. Hine. Famous photographic studies of construction workers, railroad men, factory workers and coal miners. New supplement of 18 photos on Empire State building construction. New introduction by Jonathan L. Doherty. Total of 69 photos. 63pp. 8 x 10¾. 23475-4 Pa. $4.00

THE DEPRESSION YEARS AS PHOTOGRAPHED BY ARTHUR ROTHSTEIN, Arthur Rothstein. First collection devoted entirely to the work of outstanding 1930s photographer: famous dust storm photo, ragged children, unemployed, etc. 120 photographs. Captions. 119pp. 9¼ x 10¾.
23590-4 Pa. **$5.95**

CAMERA WORK: A PICTORIAL GUIDE, Alfred Stieglitz. All 559 illustrations and plates from the most important periodical in the history of art photography, *Camera Work* (1903-17). Presented four to a page, reduced in size but still clear, in strict chronological order, with complete captions. Three indexes. Glossary. Bibliography. 176pp. 8⅜ x 11¼.
23591-2 Pa. $6.95

ALVIN LANGDON COBURN, PHOTOGRAPHER, Alvin L. Coburn. Revealing autobiography by one of greatest photographers of 20th century gives insider's version of Photo-Secession, plus comments on his own work. 77 photographs by Coburn. Edited by Helmut and Alison Gernsheim. 160pp. 8⅛ x 11.
23685-4 Pa. $6.00

NEW YORK IN THE FORTIES, Andreas Feininger. 162 brilliant photographs by the well-known photographer, formerly with *Life* magazine, show commuters, shoppers, Times Square at night, Harlem nightclub, Lower East Side, etc. Introduction and full captions by John von Hartz. 181pp. 9¼ x 10¾.
23585-8 Pa. $6.95

GREAT NEWS PHOTOS AND THE STORIES BEHIND THEM, John Faber. Dramatic volume of 140 great news photos, 1855 through 1976, and revealing stories behind them, with both historical and technical information. Hindenburg disaster, shooting of Oswald, nomination of Jimmy Carter, etc. 160pp. 8¼ x 11.
23667-6 Pa. **$6.00**

THE ART OF THE CINEMATOGRAPHER, Leonard Maltin. Survey of American cinematography history and anecdotal interviews with 5 masters—Arthur Miller, Hal Mohr, Hal Rosson, Lucien Ballard, and Conrad Hall. Very large selection of behind-the-scenes production photos. 105 photographs. Filmographies. Index. Originally *Behind the Camera.* 144pp. 8¼ x 11.
23686-2 Pa. $5.00

DESIGNS FOR THE THREE-CORNERED HAT (LE TRICORNE), Pablo Picasso. 32 fabulously rare drawings—including 31 color illustrations of costumes and accessories—for 1919 production of famous ballet. Edited by Parmenia Migel, who has written new introduction. 48pp. 9⅜ x 12¼. (Available in U.S. only)
23709-5 Pa. $5.00

NOTES OF A FILM DIRECTOR, Sergei Eisenstein. Greatest Russian filmmaker explains montage, making of *Alexander Nevsky,* aesthetics; comments on self, associates, great rivals (Chaplin), similar material. 78 illustrations. 240pp. 5⅜ x 8½.
22392-2 Pa. **$7.00**

HOLLYWOOD GLAMOUR PORTRAITS, edited by John Kobal. 145 photos capture the stars from 1926-49, the high point in portrait photography. Gable, Harlow, Bogart, Bacall, Hedy Lamarr, Marlene Dietrich, Robert Montgomery, Marlon Brando, Veronica Lake; 94 stars in all. Full background on photographers, technical aspects, much more. Total of 160pp. 8⅜ x 11¼. 23352-9 Pa. $6.95

THE NEW YORK STAGE: FAMOUS PRODUCTIONS IN PHOTOGRAPHS, edited by Stanley Appelbaum. 148 photographs from Museum of City of New York show 142 plays, 1883-1939. *Peter Pan, The Front Page, Dead End, Our Town,* O'Neill, hundreds of actors and actresses, etc. Full indexes. 154pp. 9½ x 10. 23241-7 Pa. $6.00

DIALOGUES CONCERNING TWO NEW SCIENCES, Galileo Galilei. Encompassing 30 years of experiment and thought, these dialogues deal with geometric demonstrations of fracture of solid bodies, cohesion, leverage, speed of light and sound, pendulums, falling bodies, accelerated motion, etc. 300pp. 5⅜ x 8½. 60099-8 Pa. $5.50

THE GREAT OPERA STARS IN HISTORIC PHOTOGRAPHS, edited by James Camner. 343 portraits from the 1850s to the 1940s: Tamburini, Mario, Caliapin, Jeritza, Melchior, Melba, Patti, Pinza, Schipa, Caruso, Farrar, Steber, Gobbi, and many more—270 performers in all. Index. 199pp. 8⅜ x 11¼. 23575-0 Pa. $7.50

J. S. BACH, Albert Schweitzer. Great full-length study of Bach, life, background to music, music, by foremost modern scholar. Ernest Newman translation. 650 musical examples. Total of 928pp. 5⅜ x 8½. (Available in U.S. only) 21631-4, 21632-2 Pa., Two-vol. set $12.00

COMPLETE PIANO SONATAS, Ludwig van Beethoven. All sonatas in the fine Schenker edition, with fingering, analytical material. One of best modern editions. Total of 615pp. 9 x 12. (Available in U.S. only) 23134-8, 23135-6 Pa., Two-vol. set $17.90

KEYBOARD MUSIC, J. S. Bach. Bach-Gesellschaft edition. For harpsichord, piano, other keyboard instruments. English Suites, French Suites, Six Partitas, Goldberg Variations, Two-Part Inventions, Three-Part Sinfonias. 312pp. 8⅛ x 11. (Available in U.S. only) 22360-4 Pa. $7.95

FOUR SYMPHONIES IN FULL SCORE, Franz Schubert. Schubert's four most popular symphonies: No. 4 in C Minor ("Tragic"); No. 5 in B-flat Major; No. 8 in B Minor ("Unfinished"); No. 9 in C Major ("Great"). Breitkopf & Hartel edition. Study score. 261pp. 9⅜ x 12¼. 23681-1 Pa. $8.95

THE AUTHENTIC GILBERT & SULLIVAN SONGBOOK, W. S. Gilbert, A. S. Sullivan. Largest selection available; 92 songs, uncut, original keys, in piano rendering approved by Sullivan. Favorites and lesser-known fine numbers. Edited with plot synopses by James Spero. 3 illustrations. 399pp. 9 x 12. 23482-7 Pa.$10.95

PRINCIPLES OF ORCHESTRATION, Nikolay Rimsky-Korsakov. Great classical orchestrator provides fundamentals of tonal resonance, progression of parts, voice and orchestra, tutti effects, much else in major document. 330pp. of musical excerpts. 489pp. 6½ x 9¼. 21266-1 Pa. **$7.50**

TRISTAN UND ISOLDE, Richard Wagner. Full orchestral score with complete instrumentation. Do not confuse with piano reduction. Commentary by Felix Mottl, great Wagnerian conductor and scholar. Study score. 655pp. 8⅛ x 11. 22915-7 Pa. $13.95

REQUIEM IN FULL SCORE, Giuseppe Verdi. Immensely popular with choral groups and music lovers. Republication of edition published by C. F. Peters, Leipzig, n. d. German frontmaker in English translation. Glossary. Text in Latin. Study score. 204pp. 9⅜ x 12¼.
23682-X Pa. $6.50

COMPLETE CHAMBER MUSIC FOR STRINGS, Felix Mendelssohn. All of Mendelssohn's chamber music: Octet, 2 Quintets, 6 Quartets, and Four Pieces for String Quartet. (Nothing with piano is included). Complete works edition (1874-7). Study score. 283 pp. 9⅜ x 12¼.
23679-X Pa. **$7.50**

POPULAR SONGS OF NINETEENTH-CENTURY AMERICA, edited by Richard Jackson. 64 most important songs: "Old Oaken Bucket," "Arkansas Traveler," "Yellow Rose of Texas," etc. Authentic original sheet music, full introduction and commentaries. 290pp. 9 x 12. 23270-0 Pa. **$7.95**

COLLECTED PIANO WORKS, Scott Joplin. Edited by Vera Brodsky Lawrence. Practically all of Joplin's piano works—rags, two-steps, marches, waltzes, etc., 51 works in all. Extensive introduction by Rudi Blesh. Total of 345pp. 9 x 12. 23106-2 Pa. $15.95

BASIC PRINCIPLES OF CLASSICAL BALLET, Agrippina Vaganova. Great Russian theoretician, teacher explains methods for teaching classical ballet; incorporates best from French, Italian, Russian schools. 118 illustrations. 175pp. 5⅜ x 8½. 22036-2 Pa. $2.75

CHINESE CHARACTERS, L. Wieger. Rich analysis of 2300 characters according to traditional systems into primitives. Historical-semantic analysis to phonetics (Classical Mandarin) and radicals. 820pp. 6⅛ x 9¼.
21321-8 Pa. $12.50

THE WARES OF THE MING DYNASTY, R. L. Hobson. Foremost scholar examines and illustrates many varieties of Ming (1368-1644). Famous blue and white, polychrome, lesser-known styles and shapes. 117 illustrations, 9 full color, of outstanding pieces. Total of 263pp. 6⅛ x 9¼. (Available in U.S. only) 23652-8 Pa. $6.00

AN ETYMOLOGICAL DICTIONARY OF MODERN ENGLISH, Ernest Weekley. Richest, fullest work, by foremost British lexicographer. Detailed word histories. Inexhaustible. Do not confuse this with *Concise Etymological Dictionary,* which is abridged. Total of 856pp. 6½ x 9¼.
21873-2, 21874-0 Pa., Two-vol. set $13.00

UNCLE SILAS, J. Sheridan LeFanu. Victorian Gothic mystery novel, considered by many best of period, even better than Collins or Dickens. Wonderful psychological terror. Introduction by Frederick Shroyer. 436pp. 5⅜ x 8½. 21715-9 Pa. **$6.95**

JURGEN, James Branch Cabell. The great erotic fantasy of the 1920's that delighted thousands, shocked thousands more. Full final text, Lane edition with 13 plates by Frank Pape. 346pp. 5⅜ x 8½. 23507-6 Pa. $4.50

THE CLAVERINGS, Anthony Trollope. Major novel, chronicling aspects of British Victorian society, personalities. Reprint of Cornhill serialization, 16 plates by M. Edwards; first reprint of full text. Introduction by Norman Donaldson. 412pp. 5⅜ x 8½. 23464-9 Pa. $5.00

KEPT IN THE DARK, Anthony Trollope. Unusual short novel about Victorian morality and abnormal psychology by the great English author. Probably the first American publication. Frontispiece by Sir John Millais. 92pp. 6½ x 9¼. 23609-9 Pa. $2.50

RALPH THE HEIR, Anthony Trollope. Forgotten tale of illegitimacy, inheritance. Master novel of Trollope's later years. Victorian country estates, clubs, Parliament, fox hunting, world of fully realized characters. Reprint of 1871 edition. 12 illustrations by F. A. Faser. 434pp. of text. 5⅜ x 8½. 23642-0 Pa. $6.50

YEKL and THE IMPORTED BRIDEGROOM AND OTHER STORIES OF THE NEW YORK GHETTO, Abraham Cahan. Film *Hester Street* based on *Yekl* (1896). Novel, other stories among first about Jewish immigrants of N.Y.'s East Side. Highly praised by W. D. Howells—Cahan "a new star of realism." New introduction by Bernard G. Richards. 240pp. 5⅜ x 8½. 22427-9 Pa. $3.50

THE HIGH PLACE, James Branch Cabell. Great fantasy writer's enchanting comedy of disenchantment set in 18th-century France. Considered by some critics to be even better than his famous *Jurgen*. 10 illustrations and numerous vignettes by noted fantasy artist Frank C. Pape. 320pp. 5⅜ x 8½. 23670-6 Pa. $4.00

ALICE'S ADVENTURES UNDER GROUND, Lewis Carroll. Facsimile of ms. Carroll gave Alice Liddell in 1864. Different in many ways from final Alice. Handlettered, illustrated by Carroll. Introduction by Martin Gardner. 128pp. 5⅜ x 8½. 21482-6 Pa. $2.50

FAVORITE ANDREW LANG FAIRY TALE BOOKS IN MANY COLORS, Andrew Lang. The four Lang favorites in a boxed set—the complete *Red, Green, Yellow* and *Blue* Fairy Books. 164 stories; 439 illustrations by Lancelot Speed, Henry Ford and G. P. Jacomb Hood. Total of about 1500pp. 5⅜ x 8½. 23407-X Boxed set, Pa. $16.95

HOUSEHOLD STORIES BY THE BROTHERS GRIMM. All the great Grimm stories: "Rumpelstiltskin," "Snow White," "Hansel and Gretel," etc., with 114 illustrations by Walter Crane. 269pp. 5⅜ x 8½.
21080-4 Pa. $3.50

SLEEPING BEAUTY, illustrated by Arthur Rackham. Perhaps the fullest, most delightful version ever, told by C. S. Evans. Rackham's best work. 49 illustrations. 110pp. 7⅞ x 10¾. 22756-1 Pa. $2.95

AMERICAN FAIRY TALES, L. Frank Baum. Young cowboy lassoes Father Time; dummy in Mr. Floman's department store window comes to life; and 10 other fairy tales. 41 illustrations by N. P. Hall, Harry Kennedy, Ike Morgan, and Ralph Gardner. 209pp. 5⅜ x 8½. 23643-9 Pa. $3.00

THE WONDERFUL WIZARD OF OZ, L. Frank Baum. Facsimile in full color of America's finest children's classic. Introduction by Martin Gardner. 143 illustrations by W. W. Denslow. 267pp. 5⅜ x 8½.
20691-2 Pa. $4.50

THE TALE OF PETER RABBIT, Beatrix Potter. The inimitable Peter's terrifying adventure in Mr. McGregor's garden, with all 27 wonderful, full-color Potter illustrations. 55pp. 4¼ x 5½. (Available in U.S. only)
22827-4 Pa. $1.50

THE STORY OF KING ARTHUR AND HIS KNIGHTS, Howard Pyle. Finest children's version of life of King Arthur. 48 illustrations by Pyle. 131pp. 6⅛ x 9¼. 21445-1 Pa. $5.95

CARUSO'S CARICATURES, Enrico Caruso. Great tenor's remarkable caricatures of self, fellow musicians, composers, others. Toscanini, Puccini, Farrar, etc. Impish, cutting, insightful. 473 illustrations. Preface by M. Sisca. 217pp. 8⅜ x 11¼. 23528-9 Pa. $6.95

PERSONAL NARRATIVE OF A PILGRIMAGE TO ALMADINAH AND MECCAH, Richard Burton. Great travel classic by remarkably colorful personality. Burton, disguised as a Moroccan, visited sacred shrines of Islam, narrowly escaping death. Wonderful observations of Islamic life, customs, personalities. 47 illustrations. Total of 959pp. 5⅜ x 8½.
21217-3, 21218-1 Pa., Two-vol. set $14.00

INCIDENTS OF TRAVEL IN YUCATAN, John L. Stephens. Classic (1843) exploration of jungles of Yucatan, looking for evidences of Maya civilization. Travel adventures, Mexican and Indian culture, etc. Total of 669pp. 5⅜ x 8½. 20926-1, 20927-X Pa., Two-vol. set $7.90

AMERICAN LITERARY AUTOGRAPHS FROM WASHINGTON IRVING TO HENRY JAMES, Herbert Cahoon, et al. Letters, poems, manuscripts of Hawthorne, Thoreau, Twain, Alcott, Whitman, 67 other prominent American authors. Reproductions, full transcripts and commentary. Plus checklist of all American Literary Autographs in The Pierpont Morgan Library. Printed on exceptionally high-quality paper. 136 illustrations. 212pp. 9⅛ x 12¼. 23548-3 Pa. $12.50

AN AUTOBIOGRAPHY, Margaret Sanger. Exciting personal account of hard-fought battle for woman's right to birth control, against prejudice, church, law. Foremost feminist document. 504pp. 5⅜ x 8½.
20470-7 Pa. $7.50

MY BONDAGE AND MY FREEDOM, Frederick Douglass. Born as a slave, Douglass became outspoken force in antislavery movement. The best of Douglass's autobiographies. Graphic description of slave life. Introduction by P. Foner. 464pp. 5⅜ x 8½. 22457-0 Pa. $6.50

LIVING MY LIFE, Emma Goldman. Candid, no holds barred account by foremost American anarchist: her own life, anarchist movement, famous contemporaries, ideas and their impact. Struggles and confrontations in America, plus deportation to U.S.S.R. Shocking inside account of persecution of anarchists under Lenin. 13 plates. Total of 944pp. 5⅜ x 8½.
22543-7, 22544-5 Pa., Two-vol. set $12.00

LETTERS AND NOTES ON THE MANNERS, CUSTOMS AND CONDITIONS OF THE NORTH AMERICAN INDIANS, George Catlin. Classic account of life among Plains Indians: ceremonies, hunt, warfare, etc. Dover edition reproduces for first time all original paintings. 312 plates. 572pp. of text. 6⅛ x 9¼. 22118-0, 22119-9 Pa.. Two-vol. set $12.00

THE MAYA AND THEIR NEIGHBORS, edited by Clarence L. Hay, others. Synoptic view of Maya civilization in broadest sense, together with Northern, Southern neighbors. Integrates much background, valuable detail not elsewhere. Prepared by greatest scholars: Kroeber, Morley, Thompson, Spinden, Vaillant, many others. Sometimes called Tozzer Memorial Volume. 60 illustrations, linguistic map. 634pp. 5⅜ x 8½.
23510-6 Pa. $10.00

HANDBOOK OF THE INDIANS OF CALIFORNIA, A. L. Kroeber. Foremost American anthropologist offers complete ethnographic study of each group. Monumental classic. 459 illustrations, maps. 995pp. 5⅜ x 8½.
23368-5 Pa. $13.00

SHAKTI AND SHAKTA, Arthur Avalon. First book to give clear, cohesive analysis of Shakta doctrine, Shakta ritual and Kundalini Shakti (yoga). Important work by one of world's foremost students of Shaktic and Tantric thought. 732pp. 5⅜ x 8½. (Available in U.S. only)
23645-5 Pa. $7.95

AN INTRODUCTION TO THE STUDY OF THE MAYA HIEROGLYPHS, Syvanus Griswold Morley. Classic study by one of the truly great figures in hieroglyph research. Still the best introduction for the student for reading Maya hieroglyphs. New introduction by J. Eric S. Thompson. 117 illustrations. 284pp. 5⅜ x 8½. 23108-9 Pa. $4.00

A STUDY OF MAYA ART, Herbert J. Spinden. Landmark classic interprets Maya symbolism, estimates styles, covers ceramics, architecture, murals, stone carvings as artforms. Still a basic book in area. New introduction by J. Eric Thompson. Over 750 illustrations. 341pp. 8⅜ x 11¼.
21235-1 Pa. $6.95

CATALOGUE OF DOVER BOOKS